THE FOUNDATIONS OF ARITHMETIC

Die

Grundlagen der Arithmetik

Eine logisch mathematische Untersuchung
über den Begriff der Zahl

VON

DR. G. FREGE
a.o. Professor an der Universität Jena

BRESLAU
Verlag von Wilhelm Koebner
1884

The
Foundations of Arithmetic

A logico-mathematical enquiry into the
concept of number

GOTTLOB FREGE

ENGLISH TRANSLATION BY
J. L. AUSTIN, M.A.

Second Revised Edition

NORTHWESTERN UNIVERSITY PRESS
EVANSTON · ILLINOIS · 1980

Printed in Great Britain

Translator's Note

Words and footnotes in square brackets are insertions by the translator.

The pagination is the same as in the original German edition, except that in the original the "Analysis of Contents" pages were not numbered.

Some of Frege's references and quotations, which are not always accurate, have been corrected in the translated version.

Translator's Preface to the Second Edition

Though users of the first edition of this version will perhaps not be anywhere seriously misled in doctrine, a large number of passages in it have called, and some even howled, for improvements in fidelity or lucidity. The translator's thanks are due to several readers, and in particular to Mr. P. T. Geach, for their trouble in contributing emendations and suggestions: nothing could be more welcome than more of the same.*

There is justice in the general criticism that the version is too long. Here and there it has been possible to do something to correct this, but it is too late and too difficult now to strike a fresh compromise throughout between the claims of brevity and those of naturalness and clarity. Frege is an unusually, even at times an unduly, succinct writer, and the German text must be allowed to remain the final testimony to his style.

The translations originally chosen for Frege's principal terms remain unchanged, except that *Begriffswort* has now become "concept word" instead of "general term" and *wirklich* "actual" instead of "existent." Critics of some others of these translations have perhaps not sufficiently realized that Frege's inherited philosophical vocabulary (at least as he was using it at this period) is a dated one. It is that which was Englished by his contemporaries, the "British Idealists": and they certainly used, for example, "idea" for *Vorstellung* and "proposition" for *Satz*, though not unnaturally they attached to those words meanings different from (and doubtless less clear than) those fashionable half a century later. Frege's thought cannot be reproduced accurately, nor can his terms be translated consistently, unless we are prepared to accept, even in him, something short of complete (or contemporary) sophistication.

*) *Note:* The distinguished translator died in 1960. In the 1980 impression a few amendments such as he called for have been made at the suggestion of Professor M. A. E. Dummett and Mr B. F. McGuinness.

Inhalt.

I. Meinungen einiger Schriftsteller über die Natur der arithmetischen Sätze.

Sind die Zahlformeln beweisbar?

Analysis of Contents.

I. Views of certain writers on the nature of arithmetical propositions.

Are numerical formulae provable?

II. Views of certain writers on the concept of Number.

Ist die Zahl etwas Subjectives?

Die Anzahl als Menge.

III. Meinungen über Einheit und Eins.

Drückt das Zahlwort „Ein" eine Eigenschaft von Gegenständen aus?

Sind die Einheiten einander gleich?

Is number something subjective?

Numbers as sets.

III. Views on unity and one.

Does the number word "one" express a property of objects?

Are units identical with one another?

IV. Der Begriff der Anzahl.

Jede einzelne Zahl ist ein selbständiger Gegenstand.

Um den Begriff der Anzahl zu gewinnen, muss man den Sinn einer Zahlengleichung feststellen.

Ergänzung und Bewährung unserer Definition.

IV. The concept of Number.

Every individual number is a self-subsistent object.

To obtain the concept of Number, we must fix the sense of a numerical identity.

Our definition completed and its worth proved.

Seite

Unendliche Anzahlen

V. Schluss.

Andere Zahlen.

V. Conclusion.

EINLEITUNG

Auf die Frage, was die Zahl Eins sei, oder was das Zeichen 1 bedeute, wird man meistens die Antwort erhalten: nun, ein Ding. Und wenn man dann darauf aufmerksam macht, dass der Satz

"die Zahl Eins ist ein Ding"

keine Definition ist, weil auf der einen Seite der bestimmte Artikel, auf der andern der unbestimmte steht, dass er nur besagt, die Zahl Eins gehöre zu den Dingen, aber nicht, welches Ding sie sei, so wird man vielleicht aufgefordert, sich irgendein Ding zu wählen, das man Eins nennen wolle. Wenn aber Jeder das Recht hätte, unter diesem Namen zu verstehen, was er will, so würde derselbe Satz von der Eins für Verschiedene Verschiedenes bedeuten; es gäbe keinen gemeinsamen Inhalt solcher Sätze. Einige lehnen vielleicht die Frage mit dem Hinweise darauf ab, dass auch die Bedeutung des Buchstaben a in der Arithmetik nicht angegeben werden könne; und wenn man sage: a bedeutet eine Zahl, so könne hierin derselbe Fehler gefunden werden wie in der Definition: Eins ist ein Ding. Nun ist die Ablehnung der Frage in Bezug auf a ganz gerechtfertigt: es bedeutet keine bestimmte, angebbare Zahl, sondern dient dazu, die Allgemeinheit von Sätzen auszudrücken. Wenn man für a in $a + a - a = a$ eine beliebige aber überall dieselbe Zahl

INTRODUCTION

When we ask someone what the number one is, or what the symbol 1 means,* we get as a rule the answer "Why, a thing". And if we go on to point out that the proposition
"the number one is a thing"
is not a definition, because it has the definite article on one side and the indefinite on the other, or that it only assigns the number one to the class of things, without stating which thing it is, then we shall very likely be invited to select something for ourselves—anything we please—to call one. Yet if everyone had the right to understand by this name whatever he pleased, then the same proposition about one would mean different things for different people,—such propositions would have no common content. Some, perhaps, will decline to answer the question, pointing out that it is impossible to state, either, what is meant by the letter *a*, as it is used in arithmetic; and that if we were to say "*a* means a number," this would be open to the same objection as the definition "one is a thing." Now in the case of *a* it is quite right to decline to answer: *a* does not mean some one definite number which can be specified, but serves to express the generality of general propositions. If, in $a + a - a = a$, we put for *a*

* [I have tried throughout to translate *Bedeutung* and its cognates by "meaning" and *Sinn* and its cognates by "sense", in view of the importance Frege later attached to the distinction. But it is quite evident that he attached no special significance to the words at this period.]

setzt, so erhält man immer eine wahre Gleichung. In diesem Sinne wird der Buchstabe a gebraucht. Aber bei der Eins liegt die Sache doch wesentlich anders. Können wir in der Gleichung $1 + 1 = 2$ für 1 beidemal denselben Gegenstand, etwa den Mond setzen? Vielmehr scheint es, dass wir für die erste 1 etwas Anderes wie für die zweite setzen müssen. Woran liegt es, dass hier grade das geschehen muss, was in jenem Falle ein Fehler wäre? Die Arithmetik kommt mit dem Buchstaben a allein nicht aus, sondern muss noch andere b, c u. s. w. gebrauchen, um Beziehungen zwischen verschiedenen Zahlen allgemein auszudrücken. So sollte man denken, könnte auch das Zeichen 1 nicht genügen, wenn es in ähnlicher Weise dazu diente, den Sätzen eine Allgemeinheit zu verleihen. Aber erscheint nicht die Zahl Eins als bestimmter Gegenstand mit angebbaren Eigenschaften, z. B. mit sich selbst multiplicirt unverändert zu bleiben? In diesem Sinne kann man von a keine Eigenschaften angeben; denn was von a ausgesagt wird, ist eine gemeinsame Eigenschaft der Zahlen, während $1^1 = 1$ weder vom Monde etwas aussagt, noch von der Sonne, noch von der Sahara, noch vom Pic von Teneriffa; denn was könnte der Sinn einer solchen Aussage sein?

Auf solche Fragen werden wohl auch die meisten Mathematiker keine genügende Antwort bereit haben. Ist es nun nicht für die Wissenschaft beschämend, so im Unklaren über ihren nächstliegenden und scheinbar so einfachen Gegenstand zu sein? Um so weniger wird man sagen können, was Zahl sei. Wenn ein Begriff, der einer grossen Wissenschaft zu Grunde liegt, Schwierigkeiten darbietet, so ist es doch wohl eine unabweisbare Aufgabe, ihn genauer zu untersuchen und diese Schwierigkeiten zu überwinden, besonders da es schwer gelingen möchte, über die negativen, gebrochenen, complexen Zahlen zu voller Klarheit zu kommen, solange noch die Einsicht in die Grundlage des ganzen Baues der Arithmetik mangelhaft ist.

some number, any we please but the same throughout, we get always a true identity.* This is the sense in which the letter a is used. With one, however, the position is essentially different. Can we, in the identity $1 + 1 = 2$, put for 1 in both places some one and the same object, say the Moon? On the contrary, it looks as though, whatever we put for the first 1, we must put something different for the second. Why is it that we have to do here precisely what would have been wrong in the other case? Again, arithmetic cannot get along with a alone, but has to use further letters besides (b, c and so on), in order to express in general form relations between different numbers. It would therefore be natural to suppose that the symbol 1 too, if it served in some similar way to confer generality on propositions, could not be enough by itself. Yet surely the number one looks like a definite particular object, with properties that can be specified, for example that of remaining unchanged when multiplied by itself? In this sense, a has no properties that can be specified, since whatever can be asserted of a is a common property of all numbers, whereas $1^1 = 1$ asserts nothing of the Moon, nothing of the Sun, nothing of the Sahara, nothing of the Peak of Teneriffe; for what could be the sense of any such assertion?

Questions like these catch even mathematicians for that matter, or most of them, unprepared with any satisfactory answer. Yet is it not a scandal that our science should be so unclear about the first and foremost among its objects, and one which is apparently so simple? Small hope, then, that we shall be able to say what number is. If a concept fundamental to a mighty science gives rise to difficulties, then it is surely an imperative task to investigate it more closely until those difficulties are overcome; especially as we shall hardly succeed in finally clearing up negative numbers, or fractional or complex numbers, so long as our insight into the foundation of the whole structure of arithmetic is still defective.

* [*Gleichung*. This also means, and would often be more naturally translated, "equation". But I have generally retained "identity", because this is sometimes essential and because Frege does understand equations as identities. For similar reasons I have translated *gleich* "identical", though it can mean "equal" or even merely "similar". Cp. §§ 34, 65.]

Viele werden das freilich nicht der Mühe werth achten. Dieser Begriff ist ja, wie sie meinen, in den Elementarbüchern hinreichend behandelt und damit für das ganze Leben abgethan. Wer glaubt denn über eine so einfache Sache noch etwas lernen zu können! Für so frei von jeder Schwierigkeit hält man den Begriff der positiven ganzen Zahl, dass er für Kinder wissenschaftlich erschöpfend behandelt werden könne, und dass Jeder ohne weiteres Nachdenken und ohne Bekanntschaft mit dem, was Andere gedacht haben, genau von ihm Bescheid wisse. So fehlt denn vielfach jene erste Vorbedingung des Lernens: das Wissen des Nichtwissens. Die Folge ist, dass man sich noch immer mit einer rohen Auffassung begnügt, obwohl schon Herbart*) eine richtigere gelehrt hat. Es ist betrübend und entmuthigend, dass in dieser Weise eine Erkenntniss immer wieder verloren zu gehen droht, die schon errungen war, dass so manche Arbeit vergeblich zu werden scheint, weil man im eingebildeten Reichthume nicht nöthig zu haben glaubt, sich ihre Früchte anzueignen. Auch diese Arbeit, sehe ich wohl, ist solcher Gefahr ausgesetzt. Jene Roheit der Auffassung tritt mir entgegen, wenn das Rechnen aggregatives, mechanisches Denken genannt wird**). Ich bezweifle, dass es ein solches Denken überhaupt giebt. Aggregatives Vorstellen könnte man schon eher gelten lassen; aber es ist für das Rechnen ohne Bedeutung. Das Denken ist im Wesentlichen überall dasselbe: es kommen nicht je nach dem Gegenstande verschiedene Arten von Denkgesetzen in Betracht. Die Unterschiede bestehen nur in der grösseren oder geringeren Reinheit und Unabhängigkeit von psychologischen Einflüssen und von äussern Hilfen des Denkens wie Sprache, Zahl-

*) Sämmtliche Werke, herausgegeb. von Hartenstein, Bd. X, 1 Thl. Umriss pädagogischer Vorlesungen § 252, Anm. 2: „Zwei heisst nicht zwei Dinge, sondern Verdoppelung" u. s. w.

**) K. Fischer, System der Logik und Metaphysik oder Wissenschaftslehre, 2. Aufl. § 94.

Admittedly, many people will think this not worth the trouble. Naturally, they suppose, this concept is adequately dealt with in the elementary textbooks, where the subject is settled once and for all. Who can believe that he has anything still to learn on so simple a matter? So free from all difficulty is the concept of positive whole number held to be, that an account of it fit for children can be both scientific and exhaustive; and that every schoolboy, without any further reflexion or acquaintance with what others have thought, knows all there is to know about it. The first prerequisite for learning anything is thus utterly lacking—I mean, the knowledge that we do not know. The result is that we still rest content with the crudest of views, even though since HERBART's[1] day a better doctrine has been available. It is sad and discouraging to observe how discoveries once made are always threatening to be lost again in this way, and how much work promises to have been done in vain, because we fancy ourselves so well off that we need not bother to assimilate its results. My work too, as I am well aware, is exposed to this risk. A typical crudity confronts me, when I find calculation described as "aggregative mechanical thought".[2] I doubt whether there exists any thought whatsoever answering to this description. An aggregative imagination, even, might sooner be let pass; but that has no relevance to calculation. Thought is in essentials the same everywhere: it is not true that there are different kinds of laws of thought to suit the different kinds of objects thought about. Such differences as there are consist only in this, that the thought is more pure or less pure, less dependent or more upon psychological influences and on external aids such as words or numerals, and further to some

---

[1] Collected Works, ed. Hartenstein, Vol. X, part i, *Umriss pädagogischer Vorlesungen*, § 252, n. 2: "Two does not mean two things, but doubling" etc.

[2] K. Fischer, *System der Logik und Metaphysik oder Wissenschaftslehre*, 2nd edn., §94.

zeichen und dgl., dann etwa noch in der Feinheit des Baues der Begriffe; aber grade in dieser Rücksicht möchte die Mathematik von keiner Wissenschaft, selbst der Philosophie nicht, übertroffen werden.

Man wird aus dieser Schrift ersehen können, dass auch ein scheinbar eigenthümlich mathematischer Schluss wie der von n auf n + 1 auf den allgemeinen logischen Gesetzen beruht, dass es besondrer Gesetze des aggregativen Denkens nicht bedarf. Man kann freilich die Zahlzeichen mechanisch gebrauchen, wie man papageimässig sprechen kann; aber Denken möchte das doch kaum zu nennen sein. Es ist nur möglich, nachdem durch wirkliches Denken die mathematische Zeichensprache so ausgebildet ist, dass sie, wie man sagt, für einen denkt. Dies beweist nicht, dass die Zahlen in einer besonders mechanischen Weise, etwa wie Sandhaufen aus Quarzkörnern gebildet sind. Es liegt, denke ich, im Interesse der Mathematiker einer solchen Ansicht entgegenzutreten, welche einen hauptsächlichen Gegenstand ihrer Wissenschaft und damit diese selbst herabzusetzen geeignet ist. Aber auch bei Mathematikern findet man ganz ähnliche Aussprüche. Im Gegentheil wird man dem Zahlbegriffe einen feineren Bau zuerkennen müssen als den meisten Begriffen andrer Wissenschaften, obwohl er noch einer der einfachsten arithmetischen ist.

Um nun jenen Wahn zu widerlegen, dass in Bezug auf die positiven ganzen Zahlen eigentlich gar keine Schwierigkeiten obwalten, sondern allgemeine Uebereinstimmung herrsche, schien es mir gut, einige Meinungen von Philosophen und Mathematikern über die hier in Betracht kommenden Fragen zu besprechen. Man wird sehn, wie wenig von Einklang zu finden ist, sodass geradezu entgegengesetzte Aussprüche vorkommen. Die Einen sagen z. B.: „die Einheiten sind einander gleich", die Andern halten sie für verschieden, und beide haben Gründe für ihre Behauptung, die sich nicht kurzer Hand abweisen lassen. Hierdurch suche

extent too in the finer or coarser structure of the concepts involved; but it is precisely in this respect that mathematics aspires to surpass all other sciences, even philosophy.

The present work will make it clear that even an inference like that from n to $n + 1$, which on the face of it is peculiar to mathematics, is based on the general laws of logic, and that there is no need of special laws for aggregative thought. It is possible, of course, to operate with figures mechanically, just as it is possible to speak like a parrot: but that hardly deserves the name of thought. It only becomes possible at all after the mathematical notation has, as a result of genuine thought, been so developed that it does the thinking for us, so to speak. This does not prove that numbers are formed in some peculiarly mechanical way, as sand, say, is formed out of quartz granules. In their own interests mathematicians should, I consider, combat any view of this kind, since it is calculated to lead to the disparagement of a principal object of their study, and of their science itself along with it. Yet even in the works of mathematicians are to be found expressions of exactly the same sort. The truth is quite the other way: the concept of number, as we shall be forced to recognize, has a finer structure than most of the concepts of the other sciences, even although it is still one of the simplest in arithmetic.

In order, then, to dispel this illusion that the positive whole numbers really present no difficulties at all, but that universal concord reigns about them, I have adopted the plan of criticizing some of the views put forward by mathematicians and philosophers on the questions involved. It will be seen how small is the extent of their agreement—so small, that we find one dictum precisely contradicting another. For example, some hold that "units are identical with one another," others that they are different, and each side supports its assertion with arguments that cannot be rejected out of hand. My object in this is

ich das Bedürfniss nach einer genaueren Untersuchung zu wecken. Zugleich will ich durch die vorausgeschickte Beleuchtung der von Andern ausgesprochenen Ansichten meiner eignen Auffassung den Boden ebnen, damit man sich vorweg überzeuge, dass jene andern Wege nicht zum Ziele führen, und dass meine Meinung nicht eine von vielen gleichberechtigten ist; und so hoffe ich die Frage wenigstens in der Hauptsache endgiltig zu entscheiden.

Freilich sind meine Ausführungen hierdurch wohl philosophischer geworden, als vielen Mathematikern angemessen scheinen mag; aber eine gründliche Untersuchung des Zahlbegriffes wird immer etwas philosophisch ausfallen müssen. Diese Aufgabe ist der Mathematik und Philosophie gemeinsam.

Wenn das Zusammenarbeiten dieser Wissenschaften trotz mancher Anläufe von beiden Seiten nicht ein so gedeihliches ist, wie es zu wünschen und wohl auch möglich wäre, so liegt das, wie mir scheint, an dem Ueberwiegen psychologischer Betrachtungsweisen in der Philosophie, die selbst in die Logik eindringen. Mit dieser Richtung hat die Mathematik gar keine Berührungspunkte, und daraus erklärt sich leicht die Abneigung vieler Mathematiker gegen philosophische Betrachtungen. Wenn z. B. Stricker*) die Vorstellungen der Zahlen motorisch, von Muskelgefühlen abhängig nennt, so kann der Mathematiker seine Zahlen darin nicht wiedererkennen und weiss mit einem solchen Satze nichts anzufangen. Eine Arithmetik, die auf Muskelgefühle gegründet wäre, würde gewiss recht gefühlvoll, aber auch ebenso verschwommen ausfallen wie diese Grundlage. Nein, mit Gefühlen hat die Arithmetik gar nichts zu schaffen. Ebensowenig mit innern Bildern, die aus Spuren früherer Sinneseindrücke zusammengeflossen sind. Das Schwankende und Unbestimmte, welches alle diese Gestaltungen haben, steht im starken Gegensatze zu der Bestimmtheit und

*) Studien über Association der Vorstellungen. Wien 1883.

to awaken a desire for a stricter enquiry. At the same time this preliminary examination of the views others have put forward should clear the ground for my own account, by convincing my readers in advance that these other paths do not lead to the goal, and that my opinion is not just one among many all equally tenable; and in this way I hope to settle the question finally, at least in essentials.

I realize that, as a result, I have been led to pursue arguments more philosophical than many mathematicians may approve; but any thorough investigation of the concept of number is bound always to turn out rather philosophical. It is a task which is common to mathematics and philosophy.

It may well be that the co-operation between these two sciences, in spite of many démarches from both sides, is not so flourishing as could be wished and would, for that matter, be possible. And if so, this is due in my opinion to the pre-dominance in philosophy of psychological methods of argu-ment, which have penetrated even into the field of logic. With this tendency mathematics is completely out of sympathy, and this easily accounts for the aversion to philosophical arguments felt by many mathematicians. When STRICKER,[1] for instance, calls our ideas* of numbers motor phenomena and makes them dependent on muscular sensations, no mathe-matician can recognize his numbers in such stuff or knows what on earth to make such a proposition. An arithmetic founded on muscular sensations would certainly turn out sensational enough, but also every bit as vague as its foundation. No, sensations are absolutely no concern of arithmetic. No more are mental pictures, formed from the amalgamated traces of earlier sense-impressions. All these phases of consciousness are characteristically fluctuating and indefinite, in strong con-trast to the definiteness and fixity of the concepts and objects of

[1] *Studien über Association der Vorstellungen*, Vienna 1883.

* [*Vorstellungen*. I have translated this word consistently by "idea", and cognate words by "imagine", "imagination", etc. For Frege it is a psychological term, cp. p. x^e below.]

Festigkeit der mathematischen Begriffe und Gegenstände. Es mag ja von Nutzen sein, die Vorstellungen und deren Wechsel zu betrachten, die beim mathematischen Denken vorkommen; aber die Psychologie bilde sich nicht ein, zur Begründung der Arithmetik irgendetwas beitragen zu können. Dem Mathematiker als solchem sind diese innern Bilder, ihre Entstehung und Veränderung gleichgiltig. Stricker sagt selbst, dass er sich beim Worte „Hundert" weiter nichts vorstellt als das Zeichen 100. Andere mögen sich den Buchstaben C oder sonst etwas vorstellen; geht daraus nicht hervor, dass diese innern Bilder in unserm Falle für das Wesen der Sache vollkommen gleichgiltig und zufällig sind, ebenso zufällig wie eine schwarze Tafel und ein Stück Kreide, dass sie überhaupt nicht Vorstellungen der Zahl Hundert zu heissen verdienen? Man sehe doch nicht das Wesen der Sache in solchen Vorstellungen! Man nehme nicht die Beschreibung, wie eine Vorstellung entsteht, für eine Definition und nicht die Angabe der seelischen und leiblichen Bedingungen dafür, dass uns ein Satz zum Bewusstsein kommt, für einen Beweis und verwechsele das Gedachtwerden eines Satzes nicht mit seiner Wahrheit! Man muss, wie es scheint, daran erinnern, dass ein Satz ebensowenig aufhört, wahr zu sein, wenn ich nicht mehr an ihn denke, wie die Sonne vernichtet wird, wenn ich die Augen schliesse. Sonst kommen wir noch dahin, dass man beim Beweise des pythagoräischen Lehrsatzes es nöthig findet, des Phosphorgehaltes unseres Gehirnes zu gedenken, und dass ein Astronom sich scheut, seine Schlüsse auf längst vergangene Zeiten zu erstrecken, damit man ihm nicht einꞏ wende: „du rechnest da $2 . 2 = 4$; aber die Zahlvorstellung hat ja eine Entwickelung, eine Geschichte! Man kann zweifeln, ob sie damals schon so weit war. Woher weisst du, dass in jener Vergangenheit dieser Satz schon bestand? Könnten die damals lebenden Wesen nicht den Satz $2 . 2 = 5$ gehabt haben, aus dem sich erst durch natürliche Züchtung

mathematics. It may, of course, serve some purpose to investigate the ideas and changes of ideas which occur during the course of mathematical thinking; but psychology should not imagine that it can contribute anything whatever to the foundation of arithmetic. To the mathematician as such these mental pictures, with their origins and their transformations, are immaterial. STRICKER himself states that the only idea he associates with the word "hundred" is the symbol 100. Others may have the idea of the letter C or something else; does it not follow, therefore, that these mental pictures are, so far as concerns us and the essentials of our problem, completely immaterial and incidental—as incidental as chalk and blackboard, and indeed that they do not deserve to be called ideas of the number a hundred at all? Never, then, let us suppose that the essence of the matter lies in such ideas. Never let us take a description of the origin of an idea for a definition, or an account of the mental and physical conditions on which we become conscious of a proposition for a proof of it. A proposition may be thought, and again it may be true; let us never confuse these two things. We must remind ourselves, it seems, that a proposition no more ceases to be true when I cease to think of it than the sun ceases to exist when I shut my eyes. Otherwise, in proving Pythagoras' theorem we should be reduced to allowing for the phosphorous content of the human brain; and astronomers would hesitate to draw any conclusions about the distant past, for fear of being charged with anachronism,—with reckoning twice two as four regardless of the fact that our idea of number is a product of evolution and has a history behind it. It might be doubted whether by that time it had progressed so far. How could they profess to know that the proposition $2 \times 2 = 4$ already held good in that remote epoch? Might not the creatures then extant have held the proposition $2 \times 2 = 5$, from which the proposition $2 \times 2 = 4$ was only evolved later through a process of natural selection

im Kampf ums Dasein der Satz 2 . 2 = 4 entwickelt hat,
der seinerseits vielleicht dazu bestimmt ist, auf demselben
Wege sich zu 2 . 2 = 3 fortzubilden?" Est modus in rebus,
sunt certi denique fines! Die geschichtliche Betrachtungsweise, die das Werden der Dinge zu belauschen und aus
dem Werden ihr Wesen zu erkennen sucht, hat gewiss eine
grosse Berechtigung; aber sie hat auch ihre Grenzen. Wenn
in dem beständigen Flusse aller Dinge nichts Festes, Ewiges
beharrte, würde die Erkennbarkeit der Welt aufhören und
Alles in Verwirrung stürzen. Man denkt sich, wie es
scheint, dass die Begriffe in der einzelnen Seele so entstehen,
wie die Blätter an den Bäumen und meint ihr Wesen dadurch erkennen zu können, dass man ihrer Entstehung nach-
forscht und sie aus der Natur der menschlichen Seele psychologisch zu erklären sucht. Aber diese Auffassung zieht
Alles ins Subjective und hebt, bis ans Ende verfolgt, die
Wahrheit auf. Was man Geschichte der Begriffe nennt,
ist wohl entweder eine Geschichte unserer Erkenntniss der
Begriffe oder der Bedeutungen der Wörter. Durch grosse
geistige Arbeit, die Jahrhunderte hindurch andauern kann,
gelingt es oft erst, einen Begriff in seiner Reinheit zu erkennen, ihn aus den fremden Umhüllungen herauszuschälen,
die ihn dem geistigen Auge verbargen. Was soll man nun
dazu sagen, wenn jemand, statt diese Arbeit, wo sie noch
nicht vollendet scheint, fortzusetzen, sie für nichts achtet,
in die Kinderstube geht oder sich in ältesten erdenkbaren
Entwickelungsstufen der Menschheit zurückversetzt, um dort
wie J. St. Mill etwa eine Pfefferkuchen- oder Kieselstein-
arithmetik zu entdecken! Es fehlt nur noch, dem Wohl-
geschmacke des Kuchens eine besondere Bedeutung für den
Zahlbegriff zuzuschreiben. Dies ist doch das grade Gegentheil eines vernünftigen Verfahrens und jedenfalls so unmathe-
matisch wie möglich. Kein Wunder, dass die Mathematiker
nichts davon wissen wollen! Statt eine besondere Reinheit
der Begriffe da zu finden, wo man ihrer Quelle nahe zu sein

in the struggle for existence? Why, it might even be that
2 × 2 = 4 itself is destined in the same way to develop into
2 × 2 = 3! *Est modus in rebus, sunt certi denique fines!* The
historical approach, with its aim of detecting how things
begin and of arriving from these origins at a knowledge of
their nature, is certainly perfectly legitimate; but it has also its
limitations. If everything were in continual flux, and nothing
maintained itself fixed for all time, there would no longer be
any possibility of getting to know anything about the world
and everything would be plunged in confusion. We suppose,
it would seem, that concepts sprout in the individual mind like
leaves on a tree, and we think to discover their nature by
studying their birth: we seek to define them psychologically, in
terms of the nature of the human mind. But this account makes
everything subjective, and if we follow it through to the end,
does away with truth. What is known as the history of concepts is really a history either of our knowledge of concepts or
of the meanings of words. Often it is only after immense intellectual effort, which may have continued over centuries, that
humanity at last succeeds in achieving knowledge of a concept
in its pure form, in stripping off the irrelevant accretions which
veil it from the eyes of the mind. What, then, are we to say of
those who, instead of advancing this work where it is not yet
completed, despise it, and betake themselves to the nursery,
or bury themselves in the remotest conceivable periods of
human evolution, there to discover, like JOHN STUART MILL,
some gingerbread or pebble arithmetic! It remains only to
ascribe to the flavour of the bread some special meaning for
the concept of number. A procedure like this is surely the very
reverse of rational, and as unmathematical, at any rate, as it
could well be. No wonder the mathematicians turn their
backs on it. Do the concepts, as we approach their supposed
sources, reveal themselves in peculiar purity? Not at all;

glaubt, sieht man Alles verschwommen und ungesondert wie durch einen Nebel. Es ist so, als ob jemand, um Amerika kennen zu lernen, sich in die Lage des Columbus zurückversetzen wollte, als er den ersten zweifelhaften Schimmer seines vermeintlichen Indiens erblickte. Freilich beweist ein solcher Vergleich nichts; aber er verdeutlicht hoffentlich meine Meinung Es kann ja sein, dass die Geschichte der Entdeckungen in vielen Fällen als Vorbereitung für weitere Forschungen nützlich ist; aber sie darf nicht an deren Stelle treten wollen.

Dem Mathematiker gegenüber, wäre eine Bekämpfung solcher Auffassungen wohl kaum nöthig gewesen; aber da ich auch für die Philosophen die behandelten Streitfragen möglichst zum Austrage bringen wollte, war ich genöthigt, mich auf die Psychologie ein wenig einzulassen, wenn auch nur, um ihren Einbruch in die Mathematik zurückzuweisen.

Uebrigens kommen auch in mathematischen Lehrbüchern psychologische Wendungen vor. Wenn man eine Verpflichtung fühlt, eine Definition zu geben, ohne es zu können, so will man wenigstens die Weise beschreiben, wie man zu dem betreffenden Gegenstande oder Begriffe kommt. Man erkennt diesen Fall leicht daran, dass im weitern Verlaufe nie mehr auf eine solche Erklärung zurückgegriffen wird. Für Lehrzwecke ist eine Hinführung auf die Sache auch ganz am Platze; nur sollte man sie von einer Definition immer deutlich unterscheiden. Dass auch Mathematiker Beweisgründe mit innern oder äussern Bedingungen der Führung eines Beweises verwechseln können, dafür liefert E. Schröder*) ein ergötzliches Beispiel, indem er unter der Ueberschrift: „Einziges Axiom" Folgendes darbietet: „Das gedachte Princip könnte wohl das Axiom der Inhärenz der Zeichen genannt werden. Es giebt uns die Gewissheit, dass bei allen unsern Entwicklungen und Schlussfolgerungen die Zeichen in unserer

---

*) Lehrbuch der Arithmetik und Algebra.

we see everything as through a fog, blurred and undifferentiated. It is as though everyone who wished to know about America were to try to put himself back in the position of Columbus, at the time when he caught the first dubious glimpse of his supposed India. Of course, a comparison like this proves nothing; but it should, I hope, make my point clear. It may well be that in many cases the history of earlier discoveries is a useful study, as a preparation for further researches; but it should not set up to usurp their place.

So far as mathematicians are concerned, an attack on such views would indeed scarcely have been necessary; but my treatment was designed to bring each dispute to an issue for the philosophers as well, as far as possible, so that I found myself forced to enter a little into psychology, if only to repel its invasion of mathematics.

Besides, even mathematical textbooks make use of psychological expressions. When the author feels himself obliged to give a definition, yet cannot, then he tends to give at least a description of the way in which we arrive at the object or concept concerned. These cases can easily be recognized by the fact that such explanations are never referred to again in the course of the subsequent exposition. For teaching purposes, introductory devices are certainly quite legitimate; only they should always be clearly distinguished from definitions. A delightful example of the way in which even mathematicians can confuse the grounds of proof with the mental or physical conditions to be satisfied if the proof is to be given is to be found in E. SCHRÖDER.[1] Under the heading "Special Axiom" he produces the following: "The principle I have in mind might well be called the Axiom of Symbolic Stability. It guarantees us that throughout all our arguments and deductions the symbols

[1] *Lehrbuch der Arithmetik und Algebra*, [Leipzig 1873].

Erinnerung — noch fester aber am Papiere — haften" u. s. w.

So sehr sich nun die Mathematik jede Beihilfe vonseiten der Psychologie verbitten muss, so wenig kann sie ihren engen Zusammenhang mit der Logik verleugnen. Ja, ich stimme der Ansicht derjenigen bei, die eine scharfe Trennung für unthunlich halten. Soviel wird man zugeben, dass jede Untersuchung über die Bündigkeit einer Beweisführung oder die Berechtigung einer Definition logisch sein muss. Solche Fragen sind aber gar nicht von der Mathematik abzuweisen, da nur durch ihre Beantwortung die nöthige Sicherheit erreichbar ist.

Auch in dieser Richtung gehe ich freilich etwas über das Uebliche hinaus. Die meisten Mathematiker sind bei Untersuchungen ähnlicher Art zufrieden, dem unmittelbaren Bedürfnisse genügt zu haben. Wenn sich eine Definition willig zu den Beweisen hergiebt, wenn man nirgends auf Widersprüche stösst, wenn sich Zusammenhänge zwischen scheinbar entlegnen Sachen erkennen lassen und wenn sich dadurch eine höhere Ordnung und Gesetzmässigkeit ergiebt, so pflegt man die Definition für genügend gesichert zu halten und fragt wenig nach ihrer logischen Rechtfertigung. Dies Verfahren hat jedenfalls das Gute, dass man nicht leicht das Ziel gänzlich verfehlt. Auch ich meine, dass die Definitionen sich durch ihre Fruchtbarkeit bewähren müssen, durch die Möglichkeit, Beweise mit ihnen zu führen. Aber es ist wohl zu beachten, dass die Strenge der Beweisführung ein Schein bleibt, mag auch die Schlusskette lückenlos sein, wenn die Definitionen nur nachträglich dadurch gerechtfertigt werden, dass man auf keinen Widerspruch gestossen ist. So hat man im Grunde immer nur eine erfahrungsmässige Sicherheit erlangt und muss eigentlich darauf gefasst sein, zuletzt doch noch einen Widerspruch anzutreffen, der das ganze Gebäude zum Einsturze bringt. Darum glaubte ich etwas weiter auf die allgemeinen logischen Grundlagen zurückgehn zu müssen, als vielleicht von den meisten Mathematikern für nöthig gehalten wird.

remain constant in our memory—or preferably on paper," and so on.

No less essential for mathematics than the refusal of all assistance from the direction of psychology, is the recognition of its close connexion with logic. I go so far as to agree with those who hold that it is impossible to effect any sharp separation of the two. This much everyone would allow, that any enquiry into the cogency of a proof or the justification of a definition must be a matter of logic. But such enquiries simply cannot be eliminated from mathematics, for it is only through answering them that we can attain to the necessary certainty.

In this direction too I go, certainly, further than is usual. Most mathematicians rest content, in enquiries of this kind, when they have satisfied their immediate needs. If a definition shows itself tractable when used in proofs, if no contradictions are anywhere encountered, and if connexions are revealed between matters apparently remote from one another, this leading to an advance in order and regularity, it is usual to regard the definition as sufficiently established, and few questions are asked as to its logical justification. This procedure has at least the advantage that it makes it difficult to miss the mark altogether. Even I agree that definitions must show their worth by their fruitfulness: it must be possible to use them for constructing proofs. Yet it must still be borne in mind that the rigour of the proof remains an illusion, even though no link be missing in the chain of our deductions, so long as the definitions are justified only as an afterthought, by our failing to come across any contradiction. By these methods we shall, at bottom, never have achieved more than an empirical certainty, and we must really face the possibility that we may still in the end encounter a contradiction which brings the whole edifice down in ruins. For this reason I have felt bound to go back rather further into the general logical foundations of our science than perhaps most mathematicians will consider necessary.

Als Grundsätze habe ich in dieser Untersuchung folgende festgehalten

es ist das Psychologische von dem Logischen, das Subjective von dem Objectiven scharf zu trennen;

nach der Bedeutung der Wörter muss im Satzzusammenhange, nicht in ihrer Vereinzelung gefragt werden;

der Unterschied zwischen Begriff und Gegenstand ist im Auge zu behalten.

Um das Erste zu befolgen, habe ich das Wort „Vorstellung" immer im psychologischen Sinne gebraucht und die Vorstellungen von den Begriffen und Gegenständen unterschieden. Wenn man den zweiten Grundsatz unbeachtet lässt, ist man fast genöthigt, als Bedeutung der Wörter innere Bilder oder Thaten der einzelnen Seele zu nehmen und damit auch gegen den ersten zu verstossen. Was den dritten Punkt betrifft, so ist es nur Schein, wenn man meint, einen Begriff zum Gegenstande machen zu können, ohne ihn zu verändern. Von hieraus ergiebt sich die Unhaltbarkeit einer verbreiteten formalen Theorie der Brüche, negativen Zahlen u. s. w. Wie ich die Verbesserung denke, kann ich in dieser Schrift nur andeuten. Es wird in allen diesen Fällen wie bei den positiven ganzen Zahlen darauf ankommen, den Sinn einer Gleichung festzustellen.

Meine Ergebnisse werden, denke ich, wenigstens in der Hauptsache die Zustimmung der Mathematiker finden, welche sich die Mühe nehmen, meine Gründe in Betracht zu ziehn. Sie scheinen mir in der Luft zu liegen und einzeln sind sie vielleicht schon alle wenigstens annähernd ausgesprochen worden; aber in diesem Zusammenhange mit einander möchten sie doch neu sein. Ich habe mich manchmal gewundert, dass Darstellungen, die in Einem Punkte meiner Auffassung so nahe kommen, in andern so stark abweichen.

Die Aufnahme bei den Philosophen wird je nach dem Standpunkte verschieden sein, am schlechtesten wohl bei

In the enquiry that follows, I have kept to three fundamental principles:

always to separate sharply the psychological from the
logical, the subjective from the objective;

never to ask for the meaning of a word in isolation, but
only in the context of a proposition;

never to lose sight of the distinction between concept
and object.

In compliance with the first principle, I have used the word
"idea" always in the psychological sense, and have distinguished
ideas from concepts and from objects. If the second principle
is not observed, one is almost forced to take as the meanings
of words mental pictures or acts of the individual mind, and
so to offend against the first principle as well. As to the third
point, it is a mere illusion to suppose that a concept can be
made an object without altering it. From this it follows that a
widely-held formalist theory of fractional, negative, etc., numbers is untenable. How I propose to improve upon it can be no
more than indicated in the present work. With numbers of
all these types, as with the positive whole numbers, it is a
matter of fixing the sense of an identity.

My results will, I think, at least in essentials, win the
adherence of those mathematicians who take the trouble to
attend to my arguments. They seem to me to be in the air,
and it may be that every one of them singly, or at least something very like it, has been already put forward; though
perhaps, presented as they are here in connexion with each
other, they may still be novel. I have often been astonished
at the way in which writers who on one point approach my
view so closely, on others depart from it so violently.

Their reception by philosophers will be varied, depending
on each philosopher's own position; but presumably those

jenen Empirikern, die als ursprüngliche Schlussweise nur die Induction anerkennen wollen und auch diese nicht einmal als Schlussweise, sondern als Gewöhnung. Vielleicht unterzieht Einer oder der Andere bei dieser Gelegenheit die Grundlagen seiner Erkenntnisstheorie einer erneueten Prüfung. Denen, welche etwa meine Definitionen für unnatürlich erklären möchten, gebe ich zu bedenken, dass die Frage hier nicht ist, ob natürlich, sondern ob den Kern der Sache treffend und logisch einwurfsfrei.

Ich gebe mich der Hoffnung hin, dass bei vorurtheilsloser Prüfung auch die Philosophen einiges Brauchbare in dieser Schrift finden werden.

empiricists who recognize induction as the sole original process of inference (and even that as a process not actually of inference but of habituation) will like them least. Some one or another, perhaps, will take this opportunity to examine afresh the principles of his theory of knowledge. To those who feel inclined to criticize my definitions as unnatural, I would suggest that the point here is not whether they are natural, but whether they go to the root of the matter and are logically beyond criticism.

I permit myself the hope that even the philosophers, if they examine what I have written without prejudice, will find in it something of use to them.

§ 1. Nachdem die Mathematik sich eine Zeit lang von der euklidischen Strenge entfernt hatte, kehrt sie jetzt zu ihr zurück und strebt gar über sie hinaus. In der Arithmetik war schon infolge des indischen Ursprungs vieler ihrer Verfahrungsweisen und Begriffe eine laxere Denkweise hergebracht als in der von den Griechen vornehmlich ausgebildeten Geometrie. Sie wurde durch die Erfindung der höhern Analysis nur gefördert; denn einerseits stellten sich einer strengen Behandlung dieser Lehren erhebliche, fast unbesiegliche Schwierigkeiten entgegen, deren Ueberwindung andrerseits die darauf verwendeten Anstrengungen wenig lohnen zu wollen schien. Doch hat die weitere Entwickelung immer deutlicher gelehrt, dass in der Mathematik eine blos moralische Ueberzeugung, gestützt auf viele erfolgreiche Anwendungen, nicht genügt. Für Vieles wird jetzt ein Beweis gefordert, was früher für selbstverständlich galt. Die Grenzen der Giltigkeit sind erst dadurch in manchen Fällen festgestellt worden. Die Begriffe der Function, der Stetigkeit, der Grenze, des Unendlichen haben sich einer schärferen Bestimmung bedürftig gezeigt. Das Negative und die Irrationalzahl, welche längst in die Wissenschaft aufgenommen waren, haben sich einer genaueren Prüfung ihrer Berechtigung unterwerfen müssen.

So zeigt sich überall das Bestreben, streng zu beweisen, die Giltigkeitsgrenzen genau zu ziehen und, um dies zu können, die Begriffe scharf zu fassen.

§ 1. After deserting for a time the old Euclidean standards of rigour, mathematics is now returning to them, and even making efforts to go beyond them. In arithmetic, if only because many of its methods and concepts originated in India, it has been the tradition to reason less strictly than in geometry, which was in the main developed by the Greeks. The discovery of higher analysis only served to confirm this tendency; for considerable, almost insuperable, difficulties stood in the way of any rigorous treatment of these subjects, while at the same time small reward seemed likely for the efforts expended in overcoming them. Later developments, however, have shown more and more clearly that in mathematics a mere moral conviction, supported by a mass of successful applications, is not good enough. Proof is now demanded of many things that formerly passed as self-evident. Again and again the limits to the validity of a proposition have been in this way established for the first time. The concepts of function, of continuity, of limit and of infinity have been shown to stand in need of sharper definition. Negative and irrational numbers, which had long since been admitted into science, have had to submit to a closer scrutiny of their credentials.

In all directions these same ideals can be seen at work— rigour of proof, precise delimitation of extent of validity, and as a means to this, sharp definition of concepts.

§ 2. Dieser Weg muss im weitern Verfolge auf den Begriff der Anzahl und auf die von positiven ganzen Zahlen geltenden einfachsten Sätze führen, welche die Grundlage der ganzen Arithmetik bilden. Freilich sind Zahlformeln wie $5 + 7 = 12$ und Gesetze wie das der Associativität bei der Addition durch die unzähligen Anwendungen, die tagtäglich von ihnen gemacht werden, so vielfach bestätigt, dass es fast lächerlich erscheinen kann, sie durch das Verlangen nach einem Beweise in Zweifel ziehen zu wollen. Aber es liegt im Wesen der Mathematik begründet, dass sie überall, wo ein Beweis möglich ist, ihn der Bewährung durch Induction vorzieht. Euklid beweist Vieles, was ihm jeder ohnehin zugestehen würde. Indem man sich selbst an der euklidischen Strenge nicht genügen liess, ist man auf die Untersuchungen geführt worden, welche sich an das Parallelenaxiom geknüpft haben.

So ist jene auf grösste Strenge gerichtete Bewegung schon vielfach über das zunächst gefühlte Bedürfniss hinausgegangen und dieses ist an Ausdehnung und Stärke immer gewachsen.

Der Beweis hat eben nicht nur den Zweck, die Wahrheit eines Satzes über jeden Zweifel zu erheben, sondern auch den, eine Einsicht in die Abhängigkeit der Wahrheiten von einander zu gewähren. Nachdem man sich von der Unerschütterlichkeit eines Felsblockes durch vergebliche Versuche, ihn zu bewegen, überzeugt hat, kann man ferner fragen, was ihn denn so sicher unterstütze. Je weiter man idese Untersuchungen fortsetzt, auf desto weniger Urwahrheiten führt man Alles zurück; und diese Vereinfachung ist an sich schon ein erstrebenswerthes Ziel. Vielleicht bestätigt sich auch die Hoffnung, dass man allgemeine Weisen der Begriffsbildung oder der Begründung gewinnen könne, die auch in verwickelteren Fällen verwendbar sind, indem man zum Bewusstsein bringt, was die Menschen in den einfachsten Fällen instinctiv gethan haben, und das Allgemeingiltige daraus abscheidet.

§ 2. Proceeding along these lines, we are bound eventually to come to the concept of Number* and to the simplest propositions holding of positive whole numbers, which form the foundation of the whole of arithmetic. Of course, numerical formulae like $7 + 5 = 12$ and laws like the Associative Law of Addition are so amply established by the countless applications made of them every day, that it may seem almost ridiculous to try to bring them into dispute by demanding a proof of them. But it is in the nature of mathematics always to prefer proof, where proof is possible, to any confirmation by induction. Euclid gives proofs of many things which anyone would concede him without question. And it was when men refused to be satisfied even with Euclid's standards of rigour that they were led to the enquiries set in train by the Axiom of Parallels.

Thus our movement in favour of all possible rigour has already outstripped in many directions the demand originally raised, while the demand has itself continually grown in scope and urgency.

The aim of proof is, in fact, not merely to place the truth of a proposition beyond all doubt, but also to afford us insight into the dependence of truths upon one another. After we have convinced ourselves that a boulder is immovable, by trying unsuccessfully to move it, there remains the further question, what is it that supports it so securely? The further we pursue these enquiries, the fewer become the primitive truths to which we reduce everything; and this simplification is in itself a goal worth pursuing. But there may even be justification for a further hope: if, by examining the simplest cases, we can bring to light what mankind has there done by instinct, and can extract from such procedures what is universally valid in them, may we not thus arrive at general methods for forming concepts and establishing principles which will be applicable also in more complicated cases?

---

* [*Anzahl*, i.e. cardinal number, cp. § 4 n. I have always used "Number" to translate this, and "number" for the more usual and general *Zahl*. Throughout most of the present work the distinction is not important, and Frege uses the two words almost indifferently.]

§ 3. Mich haben auch philosophische Beweggründe zu solchen Untersuchungen bestimmt. Die Fragen nach der apriorischen oder aposteriorischen, der synthetischen oder analytischen Natur der arithmetischen Wahrheiten harren hier ihrer Beantwortung. Denn, wenn auch diese Begriffe selbst der Philosophie angehören, so glaube ich doch, dass die Entscheidung nicht ohne Beihilfe der Mathematik erfolgen kann. Freilich hangt dies von dem Sinne ab, den man jenen Fragen beilegt.

Es ist kein seltener Fall, dass man zuerst den Inhalt eines Satzes gewinnt und dann auf einem andern beschwerlicheren Wege den strengen Beweis führt, durch den man oft auch die Bedingungen der Giltigkeit genauer kennen lernt. So hat man allgemein die Frage, wie wir zu dem Inhalte eines Urtheils kommen, von der zu trennen, woher wir die Berechtigung für unsere Behauptung nehmen.

Jene Unterscheidungen von apriori und aposteriori, synthetisch und analytisch betreffen nun nach meiner*) Auffassung nicht den Inhalt des Urtheils, sondern die Berechtigung zur Urtheilsfällung. Da, wo diese fehlt, fällt auch die Möglichkeit jener Eintheilung weg. Ein Irrthum apriori ist dann ein ebensolches Unding wie etwa ein blauer Begriff. Wenn man einen Satz in meinem Sinne aposteriori oder analytisch nennt, so urtheilt man nicht über die psychologischen, physiologischen und physikalischen Verhältnisse, die es möglich gemacht haben, den Inhalt des Satzes im Bewusstsein zu bilden, auch nicht darüber, wie ein Anderer vielleicht irrthümlicherweise dazu gekommen ist, ihn für wahr zu halten, sondern darüber, worauf im tiefsten Grunde die Berechtigung des Fürwahrhaltens beruht.

Dadurch wird die Frage dem Gebiete der Psychologie entrückt und dem der Mathematik zugewiesen, wenn es sich

---

*) Ich will damit natürlich nicht einen neuen Sinn hineinlegen, sondern nur das treffen, was frühere Schriftsteller, insbesondere Kant gemeint haben.

§ 3. Philosophical motives too have prompted me to enquiries of this kind. The answers to the questions raised about the nature of arithmetical truths—are they a priori or a posteriori? synthetic or analytic?—must lie in this same direction. For even though the concepts concerned may themselves belong to philosophy, yet, as I believe, no decision on these questions can be reached without assistance from mathematics—though this depends of course on the sense in which we understand them.

It not uncommonly happens that we first discover the content of a proposition, and only later give the rigorous proof of it, on other and more difficult lines; and often this same proof also reveals more precisely the conditions restricting the validity of the original proposition. In general, therefore, the question of how we arrive at the content of a judgement should be kept distinct from the other question, Whence do we derive the justification for its assertion?

Now these distinctions between a priori and a posteriori, synthetic and analytic, concern, as I see it,[1] not the content of the judgement but the justification for making the judgement. Where there is no such justification, the possibility of drawing the distinctions vanishes. An a priori error is thus as complete a nonsense as, say, a blue concept. When a proposition is called a posteriori or analytic in my sense, this is not a judgement about the conditions, psychological, physiological and physical, which have made it possible to form the content of the proposition in our consciousness; nor is it a judgement about the way in which some other man has come, perhaps erroneously, to believe it true; rather, it is a judgement about the ultimate ground upon which rests the justification for holding it to be true.

This means that the question is removed from the sphere of psychology, and assigned, if the truth concerned is a

---

[1] By this I do not, of course, mean to assign a new sense to these terms, but only to state accurately what earlier writers, KANT in particular, have meant by them.

um eine mathemathische Wahrheit handelt. Es kommt nun darauf an, den Beweis zu finden und ihn bis auf die Urwahrheiten zurückzuverfolgen. Stösst man auf diesem Wege nur auf die allgemeinen logischen Gesetze und auf Definitionen, so hat man eine analytische Wahrheit, wobei vorausgesetzt wird, dass auch die Sätze mit in Betracht gezogen werden, auf denen etwa die Zulässigkeit einer Definition beruht. Wenn es aber nicht möglich ist, den Beweis zu führen, ohne Wahrheiten zu benutzen, welche nicht allgemein logischer Natur sind, sondern sich auf ein besonderes Wissensgebiet beziehen, so ist der Satz ein synthetischer. Damit eine Wahrheit aposteriori sei, wird verlangt, dass ihr Beweis nicht ohne Berufung auf Thatsachen auskomme; d. h. auf unbeweisbare Wahrheiten ohne Allgemeinheit, die Aussagen von bestimmten Gegenständen enthalten. Ist es dagegen möglich, den Beweis ganz aus allgemeinen Gesetzen zu führen, die selber eines Beweises weder fähig noch bedürftig sind, so ist die Wahrheit apriori.*)

§ 4. Von diesen philosophischen Fragen ausgehend kommen wir zu derselben Forderung, welche unabhängig davon auf dem Gebiete der Mathematik selbst erwachsen ist: die Grundsätze der Arithmetik, wenn irgend möglich, mit grösster Strenge zu beweisen; denn nur wenn aufs sorgfältigste jede Lücke in der Schlusskette vermieden wird, kann man mit Sicherheit sagen, auf welche Urwahrheiten sich der Beweis stützt; und nur wenn man diese kennt, wird man jene Fragen beantworten können.

---

*) Wenn man überhaupt allgemeine Wahrheiten anerkennt, so muss man auch zugeben, dass es solche Urgesetze giebt, weil aus lauter einzelnen Thatsachen nichts folgt, es sei denn auf Grund eines Gesetzes. Selbst die Induction beruht auf dem allgemeinen Satze, dass dies Verfahren die Wahrheit oder doch eine Wahrscheinlichkeit für ein Gesetz begründen könne. Für den, der dies leugnet, ist die Induction nichts weiter als eine psychologische Erscheinung, eine Weise, wie Menschen zu dem Glauben an die Wahrheit eines Satzes kommen, ohne dass dieser Glaube dadurch irgendwie gerechtfertigt wäre.

mathematical one, to the sphere of mathematics. The problem becomes, in fact, that of finding the proof of the proposition, and of following it up right back to the primitive truths. If, in carrying out this process, we come only on general logical laws and on definitions, then the truth is an analytic one, bearing in mind that we must take account also of all propositions upon which the admissibility of any of the definitions depends. If, however, it is impossible to give the proof without making use of truths which are not of a general logical nature, but belong to the sphere of some special science, then the proposition is a synthetic one. For a truth to be a posteriori, it must be impossible to construct a proof of it without including an appeal to facts, i.e., to truths which cannot be proved and are not general, since they contain assertions about particular objects. But if, on the contrary, its proof can be derived exclusively from general laws, which themselves neither need nor admit of proof, then the truth is a priori.[1]

§ 4. Starting from these philosophical questions, we are led to formulate the same demand as that which had arisen independently in the sphere of mathematics, namely that the fundamental propositions of arithmetic should be proved, if in any way possible, with the utmost rigour; for only if every gap in the chain of deductions is eliminated with the greatest care can we say with certainty upon what primitive truths the proof depends; and only when these are known shall we be able to answer our original questions.

---

[1] If we recognize the existence of general truths at all, we must also admit the existence of such primitive laws, since from mere individual facts nothing follows, unless it be on the strength of a law. Induction itself depends on the general proposition that the inductive method can establish the truth of a law, or at least some probability for it. If we deny this, induction becomes nothing more than a psychological phenomenon, a procedure which induces men to believe in the truth of a proposition, without affording the slightest justification for so believing.

Wenn man nun dieser Forderung nachzukommen versucht, so gelangt man sehr bald zu Sätzen, deren Beweis solange unmöglich ist, als es nicht gelingt, darin vorkommende Begriffe in einfachere aufzulösen oder auf Allgemeineres zurückzuführen. Hier ist es nun vor allen die Anzahl, welche definirt oder als undefinirbar anerkannt werden muss. Das soll die Aufgabe dieses Buches sein.*) Von ihrer Lösung wird die Entscheidung über die Natur der arithmetischen Gesetze abhangen.

Bevor ich diese Fragen selbst angreife, will ich Einiges vorausschicken, was Fingerzeige für ihre Beantwortung geben kann. Wenn sich nämlich von andern Gesichtspunkten aus Gründe dafür ergeben, dass die Grundsätze der Arithmetik analytisch sind, so sprechen diese auch für deren Beweisbarkeit und für die Definirbarkeit des Begriffes der Anzahl. Die entgegengesetzte Wirkung werden die Gründe für die Aposteriorität dieser Wahrheiten haben. Deshalb mögen diese Streitpunkte zunächst einer vorläufigen Beleuchtung unterworfen werden.

I. Meinungen einiger Schriftsteller über die Natur der arithmetischen Sätze.

Sind die Zahlformeln beweisbar?

§ 5. Man muss die Zahlformeln, die wie $2 + 3 = 5$ von bestimmten Zahlen handeln, von den allgemeinen Gesetzen unterscheiden, die von allen ganzen Zahlen gelten.

Jene werden von einigen Philosophen**) für unbeweisbar und unmittelbar klar wie Axiome gehalten. Kant***) er-

---

*) Es wird also im Folgenden, wenn nichts weiter bemerkt wird, von keinen andern Zahlen als den positiven ganzen die Rede sein, welche auf die Frage wie viele? antworten.

**) Hobbes, Locke, Newton. Vergl. Baumann, die Lehren von Zeit, Raum und Mathematik. S. 241 u. 242, S. 365 ff., S. 475.

***) Kritik der reinen Vernunft herausgeg. v. Hartenstein. III. S. 157.

If we now try to meet this demand, we very soon come to propositions which cannot be proved so long as we do not succeed in analysing concepts which occur in them into simpler concepts or in reducing them to something of greater generality. Now here it is above all Number which has to be either defined or recognized as indefinable. This is the point which the present work is meant to settle.[1] On the outcome of this task will depend the decision as to the nature of the laws of arithmetic.

To my attack on these questions themselves I shall preface something which may give a pointer towards their answers. For suppose there should prove to be grounds from other points of view for believing that the fundamental principles of arithmetic are analytic, then these would tell also in favour of their being provable and the concept of Number definable; while any grounds for believing the same truths to be a posteriori would tell in the opposite direction. The rival theories here, therefore, may well be submitted first to a passing scrutiny.

I. Views of certain writers on the nature of arithmetical propositions.

Are numerical formulae provable?

§ 5. We must distinguish numerical formulae, such as $2 + 3 = 5$, which deal with particular numbers, from general laws, which hold good for all whole numbers.

The former are held by some philosophers[2] to be unprovable and immediately self-evident like axioms. KANT[3]

---

[1] In what follows, therefore, unless special notice is given, the only "numbers" under discussion are the positive whole numbers, which give the answer to the question "How many?".

[2] Hobbes, Locke, Newton. Cf. Baumann, *Die Lehren von Zeit, Raum und Mathematik*, [Berlin 1868, Vol. I], pp. 241–42, 365 ff., 475–76. [Hobbes, *Examinatio et Emendatio Mathematicae Hodiernae*, Amsterdam 1668, Diall. I–III, esp. I, p. 19 and III, pp. 62–63; Locke, *Essay*, Bk. IV, esp. Cap. iv, § 6 and cap. vii, §§ 6 and 10; Newton, *Arithmetica Universalis*, Vol. I, cap. i–iii, esp. iii, n. 24.]

[3] *Critique of Pure Reason*; Collected Works, ed. Hartenstein, Vol. III, p. 157 [Original edns. A 164/B205].

klärt sie für unbeweisbar und synthetisch, scheut sich aber, sie Axiome zu nennen, weil sie nicht allgemein sind, und weil ihre Zahl unendlich ist. Hankel*) nennt mit Recht diese Annahme von unendlich vielen unbeweisbaren Urwahrheiten unangemessen und paradox. Sie widerstreitet in der That dem Bedürfnisse der Vernunft nach Uebersichtlichkeit der ersten Grundlagen. Und ist es denn unmittelbar einleuchtend, dass

$$135664 + 37863 = 173527$$

ist? Nein! und eben dies führt Kant für die synthetische Natur dieser Sätze an. Es spricht aber vielmehr gegen ihre Unbeweisbarkeit; denn wie sollen sie anders eingesehen werden als durch einen Beweis, da sie unmittelbar nicht einleuchten? Kant will die Anschauung von Fingern oder Punkten zu Hilfe nehmen, wodurch er in Gefahr geräth, diese Sätze gegen seine Meinung als empirische erscheinen zu lassen; denn die Anschauung von 37863 Fingern ist doch jedenfalls keine reine. Der Ausdruck „Anschauung" scheint auch nicht recht zu passen, da schon 10 Finger durch ihre Stellungen zu einander die verschiedensten Anschauungen hervorrufen können. Haben wir denn überhaupt eine Anschauung von 135664 Fingern oder Punkten? Hätten wir sie und hätten wir eine von 37863 Fingern und eine von 173527 Fingern, so müsste die Richtigkeit unserer Gleichung sofort einleuchten, wenigstens für Finger, wenn sie unbeweisbar wäre; aber dies ist nicht der Fall.

Kant hat offenbar nur kleine Zahlen im Sinne gehabt. Dann würden die Formeln für grosse Zahlen beweisbar sein, die für kleine durch die Anschauung unmittelbar einleuchten. Aber es ist misslich, einen grundsätzlichen Unterschied zwischen kleinen und grossen Zahlen zu machen, besonders da eine scharfe Grenze nicht zu ziehen sein möchte. Wenn

*) Vorlesungen über die complexen Zahlen und ihren Functionen. S. 55.

declares them to be unprovable and synthetic, but hesitates to call them axioms because they are not general and because the number of them is infinite. HANKEL[1] justifiably calls this conception of infinitely numerous unprovable primitive truths incongruous and paradoxical. The fact is that it conflicts with one of the requirements of reason, which must be able to embrace all first principles in a survey. Besides, is it really self-evident that

$$135664 + 37863 = 173527?$$

It is not; and KANT actually urges this as an argument for holding these propositions to be synthetic. Yet it tells rather against their being unprovable; for how, if not by means of a proof, are they to be seen to be true, seeing that they are not immediately self-evident? KANT thinks he can call on our intuition of fingers or points for support, thus running the risk of making these propositions appear to be empirical, contrary to his own expressed opinion; for whatever our intuition of 37863 fingers may be, it is at least certainly not pure. Moreover, the term "intuition" seems hardly appropriate, since even 10 fingers can, in different arrangements, give rise to very different intuitions. And have we, in fact, an intuition of 135664 fingers or points at all? If we had, and if we had another of 37863 fingers and a third of 173527 fingers, then the correctness of our formula, if it were unprovable, would have to be evident right away, at least as applying to fingers; but it is not.

KANT, obviously, was thinking only of small numbers. So that for large numbers the formulae would be provable, though for small numbers they are immediately self-evident through intuition. Yet it is awkward to make a fundamental distinction between small and large numbers, especially as it would scarcely be possible to draw any sharp boundary between them. If the numerical formulae were provable

[1] *Vorlesungen über die complexen Zahlen und ihren Functionen*, p. 53.

die Zahlformeln etwa von 10 an beweisbar wären, so würde man mit Recht fragen: warum nicht von 5 an, von 2 an, von 1 an?

§ 6. Andere Philosophen und Mathematiker haben denn auch die Beweisbarkeit der Zahlformeln behauptet. Leibniz*) sagt:

„Es ist keine unmittelbare Wahrheit, dass 2 und 2 4 sind; vorausgesetzt, dass 4 bezeichnet 3 und 1. Man kann sie beweisen und zwar so:

$$\text{Definitionen: 1) 2 ist 1 und 1,}$$
$$\text{2) 3 ist 2 und 1,}$$
$$\text{3) 4 ist 3 und 1.}$$

Axiom: Wenn man Gleiches an die Stelle setzt, bleibt die Gleichung bestehen.

$$\text{Beweis: } 2 + 2 = 2 + 1 + 1 = 3 + 1 = 4.$$
$$\text{Def. 1.} \qquad \text{Def. 2.} \qquad \text{Def. 3.}$$

Also: nach dem Axiom: $2 + 2 = 4$."

Dieser Beweis scheint zunächst ganz aus Definitionen und dem angeführten Axiome aufgebaut zu sein. Auch dieses könnte in eine Definition verwandelt werden, wie es Leibniz an einem andern Orte**) selbst gethan hat. Es scheint, dass man von 1, 2, 3, 4 weiter nichts zu wissen braucht, als was in den Definitionen enthalten ist. Bei genauerer Betrachtung entdeckt man jedoch eine Lücke, die durch das Weglassen der Klammern verdeckt ist. Genauer müsste nämlich geschrieben werden:

$$2 + 2 = 2 + (1 + 1)$$
$$(2 + 1) + 1 = 3 + 1 = 4.$$

Hier fehlt der Satz

$$2 + (1 + 1) = (2 + 1) + 1,$$

der ein besonderer Fall von

$$a + (b + c) = (a + b) + c.$$

ist. Setzt man dies Gesetz voraus, so sieht man leicht, dass

---

*) Nouveaux Essais, IV. § 10. Erdm. S. 363.
**) Non inelegans specimen demonstrandi in abstractis. Erdm. S. 94.

from, say, 10 on, we should ask with justice "Why not from 5 on? or from 2 on? or from 1 on?"

§ 6. Other philosophers again, and mathematicians, have asserted that numerical formulae are actually provable. LEIBNIZ[1] says:

"It is not an immediate truth that 2 and 2 are 4; provided it be granted that 4 signifies 3 and 1. It can be proved, as follows:

> Definitions: (1) 2 is 1 and 1
> (2) 3 is 2 and 1
> (3) 4 is 3 and 1

Axiom: If equals be substituted for equals, the equality remains.*

Proof: $2 + 2 = 2 + 1 + 1$ (by Def. 1) $= 3 + 1$ (by Def. 2) $= 4$ (by Def. 3).

∴ $2 + 2 = 4$ (by the Axiom)."

This proof seems at first sight to be constructed entirely from definitions and the axiom cited. And the axiom too could be transformed into a definition, as LEIBNIZ himself does transform it in another passage.[2] It seems as though we need to know no more of 1, 2, 3 and 4 than is contained in the definitions. If we look more closely, however, we can discover a gap in the proof, which is concealed owing to the omission of the brackets. To be strictly accurate, that is, we should have to write:

$$2 + 2 = 2 + (1 + 1)$$
$$(2 + 1) + 1 = 3 + 1 = 4.$$

What is missing here is the proposition

$$2 + (1 + 1) = (2 + 1) + 1,$$

which is a special case of

$$a + (b + c) = (a + b) + c.$$

If we assume this law, it is easy to see that a similar proof can

---

[1] *Nouveaux Essais*, IV, § 10 (Erdmann edn., p. 363).

[2] *Non inelegans specimen demonstrandi in abstractis* (Erdmann edn., p. 94).

* [*Mettant des choses égales à la place, l'égalité demeure.*]

jede Formel des Einsundeins so bewiesen werden kann. Es ist dann jede Zahl aus der vorhergehenden zu definiren. In der That sehe ich nicht, wie uns etwa die Zahl 437986 angemessener gegeben werden könnte als in der leibnizischen Weise. Wir bekommen sie so, auch ohne eine Vorstellung von ihr zu haben, doch in unsere Gewalt. Die unendliche Menge der Zahlen wird durch solche Definitionen auf die Eins und die Vermehrung um eins zurückgeführt, und jede der unendlich vielen Zahlformeln kann aus einigen allgemeinen Sätzen bewiesen werden.

Dies ist auch die Meinung von H. Grassmann und H. Hankel. Jener will das Gesetz

$$a + (b + 1) = (a + b) + 1$$

durch eine Definition gewinnen, indem er sagt*):

„Wenn a und b beliebige Glieder der Grundreihe sind, so versteht man unter der Summe a + b dasjenige Glied der Grundreihe, für welches die Formel

$$a + (b + e) = a + b + e$$

gilt."

Hierbei soll e die positive Einheit bedeuten. Gegen diese Erklärung lässt sich zweierlei einwenden. Zunächst wird die Summe durch sich selbst erklärt. Wenn man noch nicht weiss, was a + b bedeuten soll, versteht man auch den Ausdruck a + (b + e) nicht. Aber dieser Einwand lässt sich vielleicht dadurch beseitigen, dass man freilich im Widerspruch mit dem Wortlaute sagt, nicht die Summe, sondern die Addition solle erklärt werden. Dann würde immer noch eingewendet werden können, dass a + b ein leeres Zeichen wäre, wenn es kein Glied der Grundreihe oder deren mehre von der verlangten Art gäbe. Dass dies nicht statthabe, setzt Grassmann einfach voraus, ohne es zu beweisen, sodass die Strenge nur scheinbar ist.

---

*) Lehrbuch der Mathematik für höhere Lehranstalten. I. Theil: Arithmetik, Stettin 1860, S. 4.

be given for every formula of addition. Every number, that means, is to be defined in terms of its predecessor. And actually I do not see how a number like 437986 could be given to us more aptly than in the way LEIBNIZ does it. Even without having any idea of it, we get it by this means at our disposal none the less. Through such definitions we reduce the whole infinite set of numbers to the number one and increase by one, and every one of the infinitely many numerical formulae can be proved from a few general propositions.

This opinion is shared by H. GRASSMANN and H. HANKEL. GRASSMANN attempts to obtain the law

$$a + (b + 1) = (a + b) + 1$$

by means of a definition, as follows[1]:

"If a and b are any arbitrary members of the basic series, then by the sum $a + b$ is to be understood that member of the basic series for which the formula

$$a + (b + e) = a + b + e$$

is valid."

e here is to be taken to mean positive unity. This definition can be criticized in two different ways. First, sum is defined in terms of itself. If we do not yet understand the meaning of $a + b$, we do not understand the expression $a + (b + e)$ either. This criticism, however, can perhaps be evaded if we say (admittedly going against the text) that what he is intending to define is not sum but addition. In that case, the criticism could still be brought that $a + b$ would be an empty symbol if there were either no member or several members of the basic series which satisfied the prescribed condition. That this does not in fact ever happen, GRASSMANN simply assumes without proof, so that the rigour of his procedure is only apparent.

[1] *Lehrbuch der Mathematik für höhere Lehranstalten*, Part i *Arithmetik*, p. 4. Stettin 1860 [= *ges. Math. u. Phys. Werke*, ed. Engel, II, i, p. 301].

9

§ 7. Man sollte denken, dass die Zahlformeln synthetisch oder analytisch, aposteriori oder apriori sind, je nachdem die allgemeinen Gesetze es sind, auf die sich ihr Beweis stützt. Dem steht jedoch die Meinung John Stuart Mill's entgegen. Zwar scheint er zunächst wie Leibniz die Wissenschaft auf Definitionen gründen zu wollen,*) da er die einzelnen Zahlen wie dieser erklärt; aber sein Vorurtheil, dass alles Wissen empirisch sei, verdirbt sofort den richtigen Gedanken wieder. Er belehrt uns nämlich,**) dass jene Definitionen keine im logischen Sinne seien, dass sie nicht nur die Bedeutung eines Ausdruckes festsetzen, sondern damit auch eine beobachtete Thatsache behaupten. Was in aller Welt mag die beobachtete oder, wie Mill auch sagt, physikalische Thatsache sein, die in der Definition der Zahl 777864 behauptet wird? Von dem ganzen Reichthume an physikalischen Thatsachen, der sich hier vor uns aufthut, nennt uns Mill nur eine einzige, die in der Definition der Zahl 3 behauptet werden soll. Sie besteht nach ihm darin, dass es Zusammenfügungen von Gegenständen giebt, welche, während sie diesen Eindruck °₀° auf die Sinne machen, in zwei Theile getrennt werden können, wie folgt : ₀₀ ₀. Wie gut doch, dass nicht Alles in der Welt niet- und nagelfest ist; dann könnten wir diese Trennung nicht vornehmen, und 2 + 1 wäre nicht 3! Wie schade, dass Mill nicht auch die physikalischen Thatsachen abgebildet hat, welche den Zahlen 0 und 1 zu Grunde liegen!

Mill fährt fort: „Nachdem dieser Satz zugegeben ist, nennen wir alle dergleichen Theile 3". Man erkennt hieraus, dass es eigentlich unrichtig ist, wenn die Uhr drei schlägt, von drei Schlägen zu sprechen, oder süss, sauer, bitter drei Geschmacksempfindungen zu nennen; ebensowenig

---

*) System der deductiven und inductiven Logik, übersetzt von J. Schiel, III. Buch, XXIV. Cap., § 5.
**) A. a. O. II. Buch, VI. Cap., § 2.

§ 7. It might well be supposed that numerical formulae would be synthetic or analytic, a posteriori or a priori, accord-ing as the general laws on which their proofs depend are so. JOHN STUART MILL, however, is of the opposite opinion. At first, indeed, he seems to mean to base the science, like LEIBNIZ, on definitions,[1] since he defines the individual numbers in the same way as LEIBNIZ; but this spark of sound sense is no sooner lit than extinguished, thanks to his pre-conception that all knowledge is empirical. He informs us, in fact,[2] that these definitions are not definitions in the logical sense; not only do they fix the meaning of a term, but they also assert along with it an observed matter of fact. But what in the world can be the observed fact, or the physical fact (to use another of MILL's expressions), which is asserted in the definition of the number 777864? Of all the whole wealth of physical facts in his apocalypse, MILL names for us only a solitary one, the one which he holds is asserted in the defini-tion of the number 3. It consists, according to him, in this, that collections of objects exist, which while they impress the senses thus, $^\circ_\circ{}^\circ$, may be separated into two parts, thus, o o o. What a mercy, then, that not everything in the world is nailed down; for if it were, we should not be able to bring off this separation, and $2 + 1$ would not be 3! What a pity that MILL did not also illustrate the physical facts underlying the numbers o and 1!

"This proposition being granted," MILL goes on, "we term all such parcels Threes." From this we can see that it is really incorrect to speak of three strokes when the clock strikes three, or to call sweet, sour and bitter three sensations

---

[1] *System of Logic*, Bk. III, cap. xxiv, § 5 (German translation by J. Schiel).
[2] Op. cit., Bk. II, cap. vi, § 2.

ist der Ausdruck „drei Auflösungsweisen einer Gleichung" zu
billigen; denn man hat niemals davon den sinnlichen Eindruck
wie von $\frac{\circ\ \circ}{\circ}$.

Nun sagt Mill: „Die Rechnungen folgen nicht aus der
Definition selbst, sondern aus der beobachteten Thatsache."
Aber wo hätte sich Leibniz in dem oben mitgetheilten
Beweise des Satzes $2 + 2 = 4$ auf die erwähnte Thatsache
berufen sollen? Mill unterlässt es die Lücke nachzuweisen,
obwohl er einen dem leibnizischen ganz entsprechenden Beweis des Satzes $5 + 2 = 7$ giebt.*) Die wirklich vorhan-
dene Lücke, die in dem Weglassen der Klammern liegt,
übersieht er wie Leibniz.

Wenn wirklich die Definition jeder einzelnen Zahl eine
besondere physikalische Thatsache behauptete, so würde man
einen Mann, der mit neunziffrigen Zahlen rechnet, nicht genug
wegen seines physikalischen Wissens bewundern können.
Vielleicht geht indessen Mill's Meinung nicht dahin, dass
alle diese Thatsachen einzeln beobachtet werden müssten,
sondern es genüge, durch Induction ein allgemeines Gesetz
abgeleitet zu haben, in dem sie sämmtlich eingeschlossen
seien. Aber man versuche, dies Gesetz auszusprechen, und
man wird finden, dass es unmöglich ist. Es reicht nicht
hin, zu sagen: es giebt grosse Sammlungen von Dingen, die
zerlegt werden können; denn damit ist nicht gesagt, dass es
so grosse Sammlungen und von der Art giebt, wie zur Definition etwa der Zahl 1000000 erfordert werden, und die
Weise der Theilung ist auch nicht genauer angegeben. Die
millsche Auffassung führt nothwendig zu der Forderung,
dass für jede Zahl eine Thatsache besonders beobachtet
werde, weil in einem allgemeinen Gesetze grade das Eigenthümliche der Zahl 1000000, das zu deren Definition nothwendig gehört, verloren gehen würde. Man dürfte nach
Mill in der That nicht setzen $1000000 = 999999 + 1$,

*) A. a. O. III. Buch, XXIV. Cap., § 5.

of taste; and equally unwarrantable is the expression "three methods of solving an equation." For none of these is a parcel which ever impresses the senses thus, $^\circ_\circ{}^\circ$.

Now according to MILL "the calculations do not follow from the definition itself but from the observed matter of fact." But at what point then, in the proof given above of the proposition $2 + 2 = 4$, ought LEIBNIZ to have appealed to the fact in question? MILL omits to point out the gap in the proof, although he gives himself a precisely analogous proof of the proposition $5 + 2 = 7$.[1] Actually, there is a gap, consisting in the omission of the brackets; but MILL overlooks this just as LEIBNIZ does.

If the definition of each individual number did really assert a special physical fact, then we should never be able sufficiently to admire, for his knowledge of nature, a man who calculates with nine-figure numbers. Meantime, perhaps MILL does not mean to go so far as to maintain that all these facts would have to be observed severally, but thinks it would be enough if we had derived through induction a general law in which they were all included together. But try to formulate this law, and it will be found impossible. It is not enough to say: "There exist large collections of things which can be split up." For this does not state that there exist collections of such a size and of such a sort as are required for, say, the number 1,000,000, nor is the manner in which they are to be divided up specified any more precisely. MILL's theory must necessarily lead to the demand that a fact should be observed specially for each number, for in a general law precisely what is peculiar to the number 1,000,000, which necessarily belongs to its definition, would be lost. On MILL's view we could actually not put $1,000,000 = 999,999 + 1$ unless

[1] Op. cit., Bk. III, cap. xxiv, § 5.

wenn man nicht grade diese eigenthümliche Weise der Zer-
legung einer Sammlung von Dingen beobachtet hätte, die
von der irgendeiner andern Zahl zukommenden verschieden ist

§ 8. Mill scheint zu meinen, dass die Definitionen
$2 = 1 + 1$, $3 = 2 + 1$, $4 = 3 + 1$ u. s. w. nicht gemacht werden dürften, ehe nicht die von ihm erwähnten
Thatsachen beobachtet wären. In der That darf man die 3
nicht als $(2 + 1)$ definiren, wenn man mit $(2 + 1)$ gar
keinen Sinn verbindet. Es fragt sich aber, ob es dazu nöthig
ist, jene Sammlung und ihre Trennung zu beobachten. Räthselhaft wäre dann die Zahl 0; denn bis jetzt hat wohl niemand 0 Kieselsteine gesehen oder getastet. Mill würde
gewiss die 0 für etwas Sinnloses, für eine blosse Redewendung
erklären; die Rechnungen mit 0 würden ein blosses Spiel
mit leeren Zeichen sein, und es wäre nur wunderbar, wie
etwas Vernünftiges dabei herauskommen könnte. Wenn aber
diese Rechnungen eine ernste Bedeutung haben, so kann auch
das Zeichen 0 selber nicht ganz sinnlos sein. Und es zeigt
sich die Möglichkeit, dass $2 + 1$ in ähnlicher Weise wie
die 0, einen Sinn auch dann noch haben könnte, wenn die
von Mill erwähnte Thatsache nicht beobachtet wäre. Wer
will in der That behaupten, dass die in der Definition einer
18ziffrigen Zahl nach Mill enthaltene Thatsache je beobachtet sei, und wer will leugnen, dass ein solches Zahlzeichen trotzdem einen Sinn habe?

Vielleicht meint man, es würden die physikalischen
Thatsachen nur für die kleineren Zahlen etwa bis 10 gebraucht, indem die übrigen aus diesen zusammengesetzt werden
könnten. Aber, wenn man 11 aus 10 und 1 blos durch Definition bilden kann, ohne die entsprechende Sammlung gesehen zu haben, so ist kein Grund, weshalb man nicht auch
die 2 aus 1 und 1 so zusammensetzen kann. Wenn die
Rechnungen mit der Zahl 11 nicht aus einer für diese be-
zeichnenden Thatsache folgen, wie kommt es, dass die Rechnungen mit der 2 sich auf die Beobachtung einer gewissen

we had observed a collection of things split up in precisely this peculiar way, different from that characteristic of any and every other number whatsoever.

§ 8. MILL seems to hold that we ought not to form the definitions $2 = 1 + 1$, $3 = 2 + 1$, $4 = 3 + 1$, and so on, unless and until the facts he refers to have been observed. It is quite true that we ought not to define 3 as $(2 + 1)$, if we attach no sense at all to $(2 + 1)$. But the question is whether, for this, it is necessary to observe his collection and its separation. If it were, the number 0 would be a puzzle; for up to now no one, I take it, has ever seen or touched 0 pebbles. MILL, of course, would explain 0 as something that has no sense, a mere manner of speaking; calculations with 0 would be a mere game, played with empty symbols, and the only wonder would be that anything rational could come of it. If, however, these calculations have a serious meaning, then the symbol 0 cannot be entirely without sense either. And the possibility suggests itself that $2 + 1$, in the same way as 0, might have a sense even without MILL's matter of fact being observed. Who is actually prepared to assert that the fact which, according to MILL, is contained in the definition of an eighteen-figure number has ever been observed, and who is prepared to deny that the symbol for such a number has, none the less, a sense?

Perhaps it is supposed that the physical facts would be used only for the smaller numbers, say up to 10, while the remaining numbers could be constructed out of these. But if we can form 11 from 10 and 1 simply by definition, without having seen the corresponding collection, then there is no reason why we should not also be able in this way to construct 2 out of 1 and 1. If calculations with the number 11 do not follow from any matter of fact uniquely characteristic of that number, how does it happen that calculations with the number

12

Sammlung und deren eigenthümlicher Trennung stützen müssen?

Man fragt vielleicht, wie die Arithmetik bestehen könne, wenn wir durch die Sinne gar keine oder nur drei Dinge unterscheiden könnten. Für unsere Kenntniss der arithmetischen Sätze und deren Anwendungen würde ein solcher Zustand gewiss etwas Missliches haben, aber auch für ihre Wahrheit? Wenn man einen Satz empirisch nennt, weil wir Beobachtungen gemacht haben müssen, um uns seines Inhalts bewusst zu werden, so gebraucht man das Wort „empirisch" nicht in dem Sinne, dass es dem „apriori" entgegengesetzt ist. Man spricht dann eine psychologische Behauptung aus, die nur den Inhalt des Satzes betrifft; ob dieser wahr sei, kommt dabei nicht in Betracht. In dem Sinne sind auch alle Geschichten Münchhausens empirisch; denn gewiss muss man mancherlei beobachtet haben, um sie erfinden zu können.

Sind die Gesetze der Arithmetik inductive Wahrheiten?

§ 9. Die bisherigen Erwägungen machen es wahrscheinlich, dass die Zahlformeln allein aus den Definitionen der einzelnen Zahlen mittels einiger allgemeinen Gesetze ableitbar sind, dass diese Definitionen beobachtete Thatsachen weder behaupten noch zu ihrer Rechtmässigkeit voraussetzen. Es kommt also darauf an, die Natur jener Gesetze zu erkennen.

Mill*) will zu seinem vorhin erwähnten Beweise der Formel $5 + 2 = 7$ den Satz „was aus Theilen zusammengesetzt ist, ist aus Theilen von diesen Theilen zusammengesetzt" benutzen. Dies hält er für einen charakteristischern Ausdruck des sonst in der Form „die Summen von Gleichem sind gleich" bekannten Satzes. Er nennt ihn inductive Wahrheit und Naturgesetz von der höchsten Ordnung. Für

*) A. a. O. III. Buch, XXIV. Cap., § 5.

2 must depend on the observation of a particular collection, separated in its own peculiar way?

It may, perhaps, be asked how arithmetic could exist, if we could distinguish nothing whatever by means of our senses, or only three things at most. Now for our knowledge, certainly, of arithmetical propositions and of their applications, such a state of affairs would be somewhat awkward—but would it affect the truth of those propositions? If we call a proposition empirical on the ground that we must have made observations in order to have become conscious of its content, then we are not using the word "empirical" in the sense in which it is opposed to "a priori". We are making a psychological statement, which concerns solely the content of the proposition; the question of its truth is not touched. In this sense, all Münchhausen's tales are empirical too; for certainly all sorts of observations must have been made before they could be invented.

Are the laws of arithmetic inductive truths?

§ 9. The considerations adduced thus far make it probable that numerical formulae can be derived from the definitions of the individual numbers alone by means of a few general laws, and that these definitions neither assert observed facts nor presuppose them for their legitimacy. Our next task, therefore, must be to ascertain the nature of the laws involved.

MILL[1] proposes to make use, for his proof (referred to above) of the formula $5 + 2 = 7$, of the principle that "Whatever is made up of parts, is made up of parts of those parts." This he holds to be an expression in more characteristic language of the principle familiar elsewhere in the form "The sums of equals are equals." He calls it an inductive truth, and a law of nature of the highest order. It is typical of the inaccuracy of

[1] Op. cit., Bk. III, cap. xxiv, § 5.

die Ungenauigkeit seiner Darstellung ist es bezeichnend, dass er diesen Satz gar nicht an der Stelle des Beweises heranzieht, wo er nach seiner Meinung unentbehrlich ist; doch scheint es, dass seine inductive Wahrheit Leibnizens Axiom vertreten soll: „wenn man Gleiches an die Stelle setzt, bleibt die Gleichung bestehen." Aber um arithmetische Wahrheiten Naturgesetze nennen zu können, legt Mill einen Sinn hinein, den sie nicht haben. Er meint z. B.*) die Gleichung $1 = 1$ könne falsch sein, weil ein Pfundstück nicht immer genau das Gewicht eines andern habe. Aber das will der Satz $1 = 1$ auch gar nicht behaupten.

Mill versteht das $+$ Zeichen so, dass dadurch die Beziehung der Theile eines physikalischen Körpers oder eines Haufens zu dem Ganzen ausgedrückt werde; aber das ist nicht der Sinn dieses Zeichens. $5 + 2 = 7$ bedeutet nicht, dass wenn man zu 5 Raumtheilen Flüssigkeit 2 Raumtheile Flüssigkeit giesst, man 7 Raumtheile Flüssigkeit erhalte, sondern dies ist eine Anwendung jenes Satzes, die nur statthaft ist, wenn nicht infolge etwa einer chemischen Einwirkung eine Volumänderung eintritt. Mill verwechselt immer Anwendungen, die man von einem arithmetischen Satze machen kann, welche oft physikalisch sind und beobachtete Thatsachen zur Voraussetzung haben, mit dem rein mathematischen Satze selber. Das Pluszeichen kann zwar in manchen Anwendungen einer Haufenbildung zu entsprechen scheinen; aber dies ist nicht seine Bedeutung; denn bei andern Anwendungen kann von Haufen, Aggregaten, dem Verhältnisse eines physikalischen Körpers zu seinen Theilen keine Rede sein, z. B. wenn man die Rechnung auf Ereignisse bezieht. Zwar kann man auch hier von Theilen sprechen; dann gebraucht man das Wort aber nicht im physikalischen oder geometrischen, sondern im logischen Sinne, wie wenn man die Ermordungen von Staatsoberhäuptern einen Theil

*) A. a. O. II. Buch, VI. Cap., § 3.

his exposition, that when he comes to the point in the proof at which, on his own view, this principle should be indispensable, he does not invoke it at all; however, it appears that his inductive truth is meant to do the work of LEIBNIZ's axiom that "If equals be substituted for equals, the equality remains." But in order to be able to call arithmetical truths laws of nature, MILL attributes to them a sense which they do not bear. For example,[1] he holds that the identity $1 = 1$ could be false, on the ground that one pound weight does not always weigh precisely the same as another. But the proposition $1 = 1$ is not intended in the least to state that it does.

MILL understands the symbol $+$ in such a way that it will serve to express the relation between the parts of a physical body or of a heap and the whole body or heap; but such is not the sense of that symbol. That if we pour 2 unit volumes of liquid into 5 unit volumes of liquid we shall have 7 unit volumes of liquid, is not the meaning of the proposition $5 + 2 = 7$, but an application of it, which only holds good provided that no alteration of the volume occurs as a result, say, of some chemical reaction. MILL always confuses the applications that can be made of an arithmetical proposition, which often are physical and do presuppose observed facts, with the pure mathematical proposition itself. The plus symbol can certainly look, in many applications, as though it corresponded to a process of heaping up; but that is not its meaning; for in other applications there can be no question of heaps or aggregates, or of the relationship between a physical body and its parts, as for example when we calculate about numbers of events. No doubt we can speak even here of "parts"; but then we are using the word not in the physical or geometrical sense, but in its logical sense, as we do when we speak of tyrannicides

---

[1] Op. cit., Bk. II, cap. vi, § 3.

der Morde überhaupt nennt. Hier hat man die logische Unterordnung. Und so entspricht auch die Addition im Allgemeinen nicht einem physikalischen Verhältnisse. Folglich können auch die allgemeinen Additionsgesetze nicht Naturgesetze sein.

§ 10. Aber sie könnten vielleicht dennoch inductive Wahrheiten sein. Wie wäre das zu denken? Von welchen Thatsachen soll man ausgehen, um sich zum Allgemeinen zu erheben? Dies können wohl nur die Zahlformeln sein. Damit verlören wir freilich den Vortheil wieder, den wir durch die Definitionen der einzelnen Zahlen gewonnen haben, und wir müssten uns nach einer andern Begründungsweise der Zahlformeln umsehen. Wenn wir uns nun auch über dies nicht ganz leichte Bedenken hinwegsetzen, so finden wir doch den Boden für die Induction ungünstig; denn hier fehlt jene Gleichförmigkeit, welche sonst diesem Verfahren eine grosse Zuverlässigkeit geben kann. Schon Leibniz*) lässt dem Philalèthe auf seine Behauptung:

„Die verschiedenen Modi der Zahl sind keiner andern Verschiedenheit fähig, als des mehr oder weniger; daher sind es einfache Modi wie die des Raumes"
antworten:

„Das kann man von der Zeit und der geraden Linie sagen, aber keinesfalls von den Figuren und noch weniger von den Zahlen, die nicht blos an Grösse verschieden, sondern auch unähnlich sind. Eine gerade Zahl kann in zwei gleiche Theile getheilt werden und nicht eine ungerade; 3 und 6 sind trianguläre Zahlen, 4 und 9 sind Quadrate, 8 ist ein Cubus u. s. f.; und dies findet bei den Zahlen noch mehr statt als bei den Figuren; denn zwei ungleiche Figuren können einander vollkommen ähnlich sein, aber niemals zwei Zahlen."

Wir haben uns zwar daran gewöhnt, die Zahlen in

---

*) Baumann, a. a. O. [Die Lehren von Raum, Zeit und Mathematik] II., S. 39; Erdm. S. 243.

as a part of murder as a whole. This is a matter of logical subordination. And in the same way addition too does not in general correspond to any physical relationship. It follows that the general laws of addition cannot, for their part, be laws of nature.

§ 10. But might they not still be inductive truths nevertheless? I do not see how that is conceivable. From what particular facts are we to take our start here, in order to advance to the general? The only available candidates for the part are the numerical formulae. Assign them to it, and of course we lose once again the advantage gained by giving our definitions of the individual numbers; we should have to cast around for some other means of establishing the numerical formulae. Even if we manage to rise superior to this misgiving too, which is not exactly easy, we shall still find the ground unfavourable for induction; for here there is none of that uniformity, which in other fields can give the method a high degree of reliability. Leibniz[1] recognized this already: for to his Philalèthe, who had asserted that

"the several modes of number are not capable of any other difference but more or less; which is why they are simple modes, like those of space,"*

he returns the answer:

"That can be said of time and of the straight line, but certainly not of the figures and still less of the numbers, which are not merely different in magnitude, but also dissimilar. An even number can be divided into two equal parts, an odd number cannot; three and six are triangular numbers, four and nine are squares, eight is a cube, and so on. And this is even more the case with the numbers than with the figures; for two unequal figures can be perfectly similar to each other, but never two numbers."

We have no doubt grown used to treating the numbers

[1] Baumann, *Die Lehren von Raum, Zeit und Mathematik*, Vol. II, p. 39 (Erdmann edn., p. 243).

* [Derived from Locke, *Essay*, Bk. II, cap. xvi, § 5.]

vielen Beziehungen als gleichartig zu betrachten; das kommt aber nur daher, weil wir eine Menge allgemeiner Sätze kennen, die von allen Zahlen gelten. Hier müssen wir uns jedoch auf den Standpunkt stellen, wo noch keiner von diesen anerkannt ist. In der That möchte es schwer sein, ein Beispiel für einen Inductionsschluss zu finden, das unserem Falle entspräche. Sonst kommt uns oft der Satz zu statten, dass jeder Ort im Raume und jeder Zeitpunkt an und für sich so gut wie jeder andere ist. Ein Erfolg muss an einem andern Orte und zu einer andern Zeit ebensogut eintreten, wenn nur die Bedingungen dieselben sind. Das fällt hier hinweg, weil die Zahlen raum- und zeitlos sind. Die Stellen in der Zahlenreihe sind nicht gleichwerthig wie die Orte des Raumes.

Die Zahlen verhalten sich auch ganz anders als die Individuen etwa einer Thierart, da sie eine durch die Natur der Sache bestimmte Rangordnung haben, da jede auf eigne Weise gebildet ist und ihre Eigenart hat, die besonders bei der o, der 1 und der 2 hervortritt. Wenn man sonst einen Satz in Bezug auf eine Gattung durch Induction begründet, hat man gewöhnlich schon eine ganze Reihe gemeinsamer Eigenschaften allein schon durch die Definition des Gattungsbegriffes. Hier hält es schwer, nur eine einzige zu finden, die nicht selbst erst nachzuweisen wäre.

Am leichtesten möchte sich unser Fall noch mit folgendem vergleichen lassen. Man habe in einem Bohrloche eine mit der Tiefe regelmässig zunehmende Temperatur bemerkt; man habe bisher sehr verschiedene Gesteinsschichten angetroffen. Es ist dann offenbar aus den Beobachtungen, die man an diesem Bohrloche gemacht hat, allein nichts über die Beschaffenheit der tiefern Schichten zu schliessen, und ob die Regelmässigkeit der Temperaturvertheilung sich weiter bewähren würde, muss dahingestellt bleiben. Unter den Begriff „was bei fortgesetztem Bohren angetroffen wird" fällt zwar das bisher Beobachtete wie das Tieferliegende;

in many contexts as all of the same sort, but that is only because we know a set of general propositions which hold for all numbers. For the present purpose, however, we must put ourselves in the position that none of these has yet been discovered. The fact is that it would be difficult to find an example of an inductive inference to parallel our present case. In ordinary inductions we often make good use of the proposition that every position in space and every moment in time is as good in itself as every other. Our results must hold good for any other place and any other time, provided only that the conditions are the same. But in the case of the numbers this does not apply, since they are not in space or time. Position in the number series is not a matter of indifference like position in space.

The numbers, moreover, are related to one another quite differently from the way in which the individual specimens of, say, a species of animal are. It is in their nature to be arranged in a fixed, definite order of precedence; and each one is formed in its own special way and has its own unique peculiarities, which are specially prominent in the cases of 0, 1 and 2. Elsewhere when we establish by induction a proposition about a species, we are ordinarily in possession already, merely from the definition of the concept of the species, of a whole series of its common properties. But with the numbers we have difficulty in finding even a single common property which has not actually to be first proved common.

The following is perhaps the case with which our putative induction might most easily be compared. Suppose we have noticed that in a borehole the temperature increases regularly with the depth; and suppose we have so far encountered a wide variety of differing rock strata. Here it is obvious that we cannot, simply on the strength of the observations made at this borehole, infer anything whatever as to the nature of the strata at deeper levels, and that any answer to the question, whether the regular distribution of temperature would continue to hold good lower down, would be premature. There is, it is true, a concept, that of "whatever you come to by going on boring," under which fall both the strata so far observed and those at lower levels alike; but that is of little assistance

aber das kann hier wenig nützen. Ebenso wenig wird es uns bei den Zahlen nützen, dass sie sämmtlich unter den Begriff „was man durch fortgesetzte Vermehrung um eins erhält" fallen. Man kann eine Verschiedenheit der beiden Fälle darin finden, dass die Schichten nur angetroffen werden, die Zahlen aber durch die fortgesetzte Vermehrung um eins geradezu geschaffen und ihrem ganzen Wesen nach bestimmt werden. Dies kann nur heissen, dass man aus der Weise, wie eine Zahl, z. B. 8, durch Vermehrung um 1 entstanden ist, alle ihre Eigenschaften ableiten kann. Damit giebt man im Grunde zu, dass die Eigenschaften der Zahlen aus ihren Definitionen folgen, und es eröffnet sich die Möglichkeit, die allgemeinen Gesetze der Zahlen aus der allen gemeinsamen Entstehungsweise zu beweisen, während die besondern Eigenschaften der einzelnen aus der besondern Weise zu folgern wären, wie sie durch fortgesetzte Vermehrung um eins gebildet sind. So kann man auch, was bei den Erdschichten, schon durch die Tiefe allein bestimmt ist, in der sie getroffen werden, also ihre Lagenverhältnisse, eben daraus schliessen, ohne dass man die Induction nöthig hätte; was aber nicht dadurch bestimmt ist, kann auch die Induction nicht lehren.

Vermuthlich kann das Verfahren der Induction selbst nur mittels allgemeiner Sätze der Arithmetik gerechtfertigt werden, wenn man darunter nicht eine blosse Gewöhnung versteht. Diese hat nämlich durchaus keine wahrheitverbürgende Kraft. Während das wissenschaftliche Verfahren nach objectiven Maasstäben bald in einer einzigen Bestätigung eine hohe Wahrscheinlichkeit begründet findet, bald tausendfaches Eintreffen fast für werthlos erachtet, wird die Gewöhnung durch Zahl und Stärke der Eindrücke und subjective Verhältnisse bestimmt, die keinerlei Recht haben, auf das Urtheil Einfluss zu üben. Die Induction muss sich auf die Lehre von der Wahrscheinlichkeit stützen, weil sie einen Satz nie mehr als wahrscheinlich machen kann. Wie

to us here. And equally, it will be no help to us to learn in the case of the numbers that these all fall together under the concept of "whatever you get by going on increasing by one." It is possible to draw a distinction between the two cases, on the ground that the strata are things we simply encounter, whereas the numbers are literally created, and determined in their whole natures, by the process of continually increasing by one. Now this can only mean that from the way in which a number, say 8, is generated through increasing by one all its properties can be deduced. But this is in principle to grant that the properties of numbers follow from their definitions, and to open up the possibility that we might prove the general laws of numbers from the method of generation which is common to them all, while deducing the special properties of the individual numbers from the special way in which, through the process of continually increasing by one, each one is formed. In the same way in the geological case too, we can deduce everything that is determined simply and solely by the depth at which a stratum is encountered, namely its spatial position relative to anything else, from the depth itself, without having any need of induction; but whatever is not so determined, cannot be learned by induction either.

The procedure of induction, we may surmise, can itself be justified only by means of general propositions of arithmetic —unless we understand by induction a mere process of habituation, in which case it has of course absolutely no power whatever of leading to the discovery of truth. The procedure of the sciences, with its objective standards, will at times find a high probability established by a single confirmatory instance, while at others it will dismiss a thousand as almost worthless; whereas our habits are determined by the number and strength of the impressions we receive and by subjective circumstances, which have no sort of right at all to influence our judgement. Induction [then, properly understood,] must base itself on the theory of probability, since it can never render a proposition more than probable. But how probability

diese Lehre aber ohne Voraussetzung arithmetischer Gesetze entwickelt werden könne, ist nicht abzusehen.

§ 11. Leibniz*) meint dagegen, dass die nothwendigen Wahrheiten, wie man solche in der Arithmetik findet, Principien haben müssen, deren Beweis nicht von den Beispielen und also nicht von dem Zeugnisse der Sinne abhangt, wiewohl ohne die Sinne sich niemand hätte einfallen lassen, daran zu denken. „Die ganze Arithmetik ist uns eingeboren und in uns auf virtuelle Weise." Wie er den Ausdruck „eingeboren" meint, verdeutlicht eine andere Stelle**): „Es ist nicht wahr, dass alles, was man lernt, nicht eingeboren sei; — die Wahrheiten der Zahlen sind in uns, und nichtsdestoweniger lernt man sie, sei es, indem man sie aus ihrer Quelle zieht, wenn man sie auf beweisende Art lernt (was eben zeigt, dass sie eingeboren sind), sei es . . .".

Sind die Gesetze der Arithmetik synthetisch apriori oder analytisch?

§ 12. Wenn man den Gegensatz von analytisch und synthetisch hinzunimmt, ergeben sich vier Combinationen, von denen jedoch eine, nämlich

analytisch aposteriori

ausfällt. Wenn man sich mit Mill für aposteriori entschieden hat, bleibt also keine Wahl, sodass für uns nur noch die Möglichkeiten

synthetisch apriori

und

analytisch

zu erwägen bleiben. Für die erstere entscheidet sich Kant.

---

*) Baumann a. a. O. Bd. II. S. 13 u. 14; Erdm. S. 195, S. 208 u. 209.
**) Baumann a. a. O. Bd. II., S. 38; Erdm. S. 212.

theory could possibly be developed without presupposing arithmetical laws is beyond comprehension.

§ 11. LEIBNIZ[1] holds the opposite view, that the necessary truths, such as are found in arithmetic, must have principles whose proof does not depend on examples and therefore not on the evidence of the senses, though doubtless without the senses it would have occurred to no one to think of them. "The whole of arithmetic is innate and is in virtual fashion in us." What he means by the term "innate" is explained by another passage, where he denies "that *Everything we learn is not innate.* The truths of number are in us and yet we still learn them, whether it be by drawing them forth from their source when learning them by demonstration (which shows them to be innate), or whether it be . . .".

Are the laws of arithmetic synthetic a priori or analytic?

§ 12. If we now bring in the other antithesis between analytic and synthetic, there result four possible combinations, of which however one, viz.

analytic a posteriori

can be eliminated. Those who have decided with MILL in favour of a posteriori have therefore no second choice, so that there remain only two possibilities for us still to examine, viz.

synthetic a priori

and

analytic.

KANT declares for the former. In that case, there is no

[1] Baumann, op. cit., Vol. II, pp. 13-14 (Erdmann edn., pp. 195, 208-9).
[2] Baumann, op. cit., Vol. II, p. 38 (Erdmann edn., p. 212).

18

In diesem Falle bleibt wohl nichts übrig, als eine reine Anschauung als letzten Erkenntnissgrund anzurufen, obwohl hier schwer zu sagen ist, ob es eine räumliche oder zeitliche ist, oder welche es sonst sein mag. Baumann*) stimmt Kant, wenngleich mit etwas anderer Begründung, bei. Auch nach Lipschitz**) fliessen die Sätze, welche die Unabhängigkeit der Anzahl von der Art des Zählens und die Vertauschbarkeit und Gruppirbarkeit der Summanden behaupten, aus der inneren Anschauung. Hankel***) gründet die Lehre von den reellen Zahlen auf drei Grundsätze, denen er den Charakter der notiones communes zuschreibt: „Sie werden durch Explication vollkommen evident, gelten für alle Grössengebiete nach der reinen Anschauung der Grösse und können, ohne ihren Charakter einzubüssen, in Definitionen verwandelt werden, indem man sagt: Unter der Addition von Grössen versteht man eine Operation, welche diesen Sätzen genügt." In der letzten Behauptung liegt eine Unklarheit. Vielleicht kann man die Definition machen; aber sie kann keinen Ersatz für jene Grundsätze bilden; denn bei der Anwendung würde es sich immer darum handeln: sind die Anzahlen Grössen, und ist das, was man Addition der Anzahlen zu nennen pflegt, Addition im Sinne dieser Definition? Und zur Beantwortung müsste man jene Sätze von den Anzahlen schon kennen. Ferner erregt der Ausdruck „reine Anschauung der Grösse" Anstoss. Wenn man erwägt, was alles Grösse genannt wird: Anzahlen, Längen, Flächeninhalte, Volumina, Winkel, Krümmungen, Massen, Geschwindigkeiten, Kräfte, Lichtstärken, galvanische Stromstärken u. s. f., so ist wohl zu verstehen, wie man dies einem Grössenbegriffe unterordnen kann; aber der Ausdruck „Anschauung der Grösse" und gar „reine An-

---

*) A. a. O. Bd. II., S. 669.
**) Lehrbuch der Analysis, Bd. I., S. 1.
***) Theorie der complexen Zahlensysteme, S. 54 u. 55.

alternative but to invoke a pure intuition as the ultimate ground of our knowledge of such judgements, hard though it is to say of this whether it is spatial or temporal, or whatever else it may be. BAUMANN[1] agrees with KANT, although for rather different reasons. LIPSCHITZ,[2] too, holds that certain propositions, namely that which asserts that Number is independent of the method of numbering and also the Commutative and Associative Laws of Addition, are derived from inner intuition. HANKEL[3] bases the theory of real numbers on three fundamental propositions, to which he ascribes the character of "common notions" (*notiones communes*): "Once expounded they are perfectly self-evident; they are valid for magnitudes in every field, as vouched for by our pure intuition of magnitude; and they can without losing their character be transformed into definitions, simply by defining the addition of magnitudes as an operation which satisfies them." In the last statement here there is an obscurity. The definition can perhaps be constructed, but it will not do as a substitute for the original propositions; for in seeking to apply it the question would always arise: Are Numbers magnitudes, and is what we ordinarily call addition of Numbers addition in the sense of this definition? And to answer it, we should need to know already his original propositions about Numbers. Moreover, the expression "pure intuition of magnitude" gives us pause. If we consider all the different things that are called magnitudes: Numbers, lengths, areas, volumes, angles, curvatures, masses, velocities, forces, illuminations, electric currents, and so forth, we can quite well understand how they can all be brought under the single *concept* of magnitude; but the term "intuition of magnitude," and still worse "pure intuition of

---

[1] Op. cit., Vol. II, p. 669.

[2] *Lehrbuch der Analysis*, Vol. I, p. 1 [Bonn 1877].

[3] *Theorie der complexen Zahlensysteme*, pp. 54–55.

schauung der Grösse" kann nicht als zutreffend anerkannt werden. Ich kann nicht einmal eine Anschauung von 100000 zugeben, noch viel weniger von Zahl im Allgemeinen oder gar von Grösse im Allgemeinen. Man beruft sich zu leicht auf innere Anschauung, wenn man keinen andern Grund anzugeben vermag. Aber man sollte dabei den Sinn des Wortes „Anschauung" doch nicht ganz aus dem Auge verlieren.

Kant definirt in der Logik (ed. Hartenstein, VIII, S. 88):

„Die Anschauung ist eine einzelne Vorstellung (repraesentatio singularis), der Begriff eine allgemeine (repraesentatio per notas communes) oder reflectirte Vorstellung (repraesentatio discursiva)."

Hier kommt die Beziehung zur Sinnlichkeit gar nicht zum Ausdrucke, die doch in der transcendentalen Aesthetik hinzugedacht wird, und ohne welche die Anschauung nicht als Erkenntnissprincip für die synthetischen Urtheile apriori dienen kann. In der Kr. d. r. V. (ed. Hartenstein III, S. 55) heisst es:

„Vermittelst der Sinnlichkeit also werden uns Gegenstände gegeben und sie allein liefert uns Anschauungen."

Der Sinn unseres Wortes in der Logik ist demnach ein weiterer als in der trancendentalen Aesthetik. Im logischen Sinne könnte man vielleicht 100000 eine Anschauung nennen; denn ein allgemeiner Begriff ist es nicht. Aber in diesem Sinne genommen, kann die Anschauung nicht zur Begründung der arithmetischen Gesetze dienen.

§ 13. Ueberhaupt wird es gut sein, die Verwandtschaft mit der Geometrie nicht zu überschätzen. Ich habe schon eine leibnizische Stelle dagegen angeführt. Ein geometrischer Punkt für sich betrachtet, ist von irgendeinem andern gar nicht zu unterscheiden; dasselbe gilt von Geraden und Ebenen. Erst wenn mehre Punkte, Gerade, Ebenen in einer Anschauung gleichzeitig aufgefasst werden, unterscheidet man sie. Wenn in der Geometrie allgemeine Sätze aus der

magnitude", cannot be admitted as appropriate. I cannot even allow an intuition of 100,000, far less of number in general, not to mention magnitude in general. We are all too ready to invoke inner intuition, whenever we cannot produce any other ground of knowledge. But we have no business, in doing so, to lose sight altogether of the sense of the word "intuition".

KANT in his *Logic* (ed. Hartenstein, vol. VIII, p. 88) defines it as follows:

"An intuition is an *individual* idea (REPRÆSENTATIO SINGULARIS), a concept is a *general* idea (REPRÆSENTATIO PER NOTAS COMMUNES) or an idea *of reflexion* (REPRÆSENTATIO DISCURSIVA)."

Here there is absolutely no mention of any connexion with sensibility, which is, however, included in the notion of intuition in the *Transcendental Aesthetic*, and without which intuition cannot serve as the principle of our knowledge of synthetic a priori judgements. In the *Critique of Pure Reason* (ed. Hartenstein, vol. III, p. 55)* we read:

"It is therefore through the medium of sensibility that objects are *given* to us and it alone provides us with *intuitions*."

It follows that the sense of the word "intuition" is wider in the *Logic* than in the *Transcendental Aesthetic*. In the sense of the *Logic*, we might perhaps be able to call 100,000 an intuition; for it is not a general concept anyhow. But an intuition in this sense cannot serve as the ground of our knowledge of the laws of arithmetic.

§ 13. We shall do well in general not to overestimate the extent to which arithmetic is akin to geometry. I have already quoted a warning to this effect from LEIBNIZ. One geometrical point, considered by itself, cannot be distinguished in any way from any other; the same applies to lines and planes. Only when several points, or lines or planes, are included together in a single intuition, do we distinguish them. In geometry, therefore, it is quite intelligible that general pro-

---

* [Original edns., A19/B33]

Anschauung gewonnen werden, so ist das daraus erklärlich, dass die angeschauten Punkte, Geraden, Ebenen eigentlich gar keine besondern sind und daher als Vertreter ihrer ganzen Gattung gelten können. Anders liegt die Sache bei den Zahlen: jede hat ihre Eigenthümlichkeit. Inwiefern eine bestimmte Zahl alle andern vertreten kann, und wo ihre Besonderheit sich geltend macht, ist ohne Weiteres nicht zu sagen.

§ 14. Auch die Vergleichung der Wahrheiten in Bezug auf das von ihnen beherrschte Gebiet spricht gegen die empirische und synthetische Natur der arithmetischen Gesetze.

Die Erfahrungssätze gelten für die physische oder psychologische Wirklichkeit, die geometrischen Wahrheiten beherrschen das Gebiet des räumlich Anschaulichen, mag es nun Wirklichkeit oder Erzeugniss der Einbildungskraft sein. Die tollsten Fieberphantasien, die kühnsten Erfindungen der Sage und der Dichter, welche Thiere reden, Gestirne stille stehen lassen, aus Steinen Menschen und aus Menschen Bäume machen, und lehren, wie man sich am eignen Schopfe aus dem Sumpfe zieht, sie sind doch, sofern sie anschaulich bleiben, an die Axiome der Geometrie gebunden. Von diesen kann nur das begriffliche Denken in gewisser Weise loskommen, wenn es etwa einen Raum von vier Dimensionen oder von positivem Krümmungsmaasse annimmt. Solche Betrachtungen sind durchaus nicht unnütz; aber sie verlassen ganz den Boden der Anschauung. Wenn man diese auch dabei zu Hilfe nimmt, so ist es doch immer die Anschauung des euklidischen Raumes, des einzigen, von dessen Gebilden wir eine haben. Sie wird dann nur nicht so, wie sie ist, sondern symbolisch für etwas anderes genommen; man nennt z. B. gerade oder eben, was man doch als Krummes anschaut. Für das begriffliche Denken kann man immerhin von diesem oder jenem geometrischen Axiome das Gegentheil annehmen, ohne dass man in Widersprüche mit sich selbst verwickelt wird, wenn man Schlussfolgerungen

positions should be derived from intuition; the points or lines or planes which we intuite are not really particular at all, which is what enables them to stand as representatives of the whole of their kind. But with the numbers it is different; each number has its own peculiarities. To what extent a given particular number can represent all the others, and at what point its own special character comes into play, cannot be laid down generally in advance.

§ 14. If, again, we compare the various kinds of truths in respect of the domains that they govern, the comparison tells once more against the supposed empirical and synthetic character of arithmetical laws.

Empirical propositions hold good of what is physically or psychologically actual, the truths of geometry govern all that is spatially intuitable, whether actual or product of our fancy. The wildest visions of delirium, the boldest inventions of legend and poetry, where animals speak and stars stand still, where men are turned to stone and trees turn into men, where the drowning haul themselves up out of swamps by their own topknots—all these remain, so long as they remain intuitable, still subject to the axioms of geometry. Conceptual thought alone can after a fashion shake off this yoke, when it assumes, say, a space of four dimensions or positive curvature. To study such conceptions is not useless by any means; but it is to leave the ground of intuition entirely behind. If we do make use of intuition even here, as an aid, it is still the same old intuition of Euclidean space, the only one whose structures we can intuit. Only then the intuition is not taken at its face value, but as symbolic of something else; for example, we call straight or plane what we actually intuite as curved. For purposes of conceptual thought we can always assume the contrary of some one or other of the geometrical axioms, without involving ourselves in any self-contradictions when we proceed to our deductions, despite

aus solchen der Anschauung widerstreitenden Annahmen zieht. Diese Möglichkeit zeigt, dass die geometrischen Axiome von einander und von den logischen Urgesetzen unabhängig, also synthetisch sind. Kann man dasselbe von den Grundsätzen der Zahlenwissenschaft sagen? Stürzt nicht alles in Verwirrung, wenn man einen von diesen leugnen wollte? Wäre dann noch Denken möglich? Liegt nicht der Grund der Arithmetik tiefer als der alles Erfahrungswissens, tiefer selbst als der der Geometrie? Die arithmetischen Wahrheiten beherrschen das Gebiet des Zählbaren. Dies ist das umfassendste; denn nicht nur das Wirkliche, nicht nur das Anschauliche gehört ihm an, sondern alles Denkbare. Sollten also nicht die Gesetze der Zahlen mit denen des Denkens in der innigsten Verbindung stehen?

§ 15. Dass Leibnizens Aussprüche sich nur zu Gunsten der analytischen Natur der Zahlgesetze deuten lassen, ist vorauszusehen, da für ihn das Apriori mit dem Analytischen zusammenfällt. So sagt er*), dass die Algebra ihre Vortheile einer viel höhern Kunst, nämlich der wahren Logik entlehne. An einer andern Stelle**) vergleicht er die nothwendigen und zufälligen Wahrheiten mit den commensurabeln und incommensurabeln Grössen und meint, dass bei nothwendigen Wahrheiten ein Beweis oder eine Zurückführung auf Identitäten möglich sei. Doch diese Aeusserungen verlieren dadurch an Gewicht, dass Leibniz dazu neigt, alle Wahrheiten als beweisbar anzusehen***): „. . . . dass jede Wahrheit ihren apriorischen, aus dem Begriff der Termini gezogenen Beweis hat, wiewohl es nicht immer in unserer Macht steht, zu dieser Analyse zu kommen." Der Vergleich mit der Commensurabilität und Incommensurabilität richtet freilich doch wieder eine für uns wenigstens unüber-

---

*) Baumann a. a. O. Bd. II., S. 56; Erdm. S. 424.
**) Baumann a. a. O. Bd. II., S. 57; Erdm. S. 83.
***) Baumann a. a. O. Bd. II., S. 57; Pertz, II., S. 55.

the conflict between our assumptions and our intuition. The fact that this is possible shows that the axioms of geometry are independent of one another and of the primitive laws of logic, and consequently are synthetic. Can the same be said of the fundamental propositions of the science of number? Here, we have only to try denying any one of them, and complete confusion ensues. Even to think at all seems no longer possible. The basis of arithmetic lies deeper, it seems, than that of any of the empirical sciences, and even than that of geometry. The truths of arithmetic govern all that is numerable. This is the widest domain of all; for to it belongs not only the actual, not only the intuitable, but everything thinkable. Should not the laws of number, then, be connected very intimately with the laws of thought?

§ 15. Statements in LEIBNIZ can only be taken to mean that the laws of number are analytic, as was to be expected, since for him the a priori coincides with the analytic. Thus he declares[1] that the benefits of algebra are due to its borrowings from a far superior science, that of the true logic. In another passage[2] he compares necessary and contingent truths to commensurable and incommensurable magnitudes, and maintains that in the case of necessary truths a proof or reduction to identities* is possible. However, these declarations lose some of their force in view of LEIBNIZ's[3] inclination to regard all truths as provable: "Every truth", he says, "has its proof a priori derived from the concept of the terms, notwithstanding it does not always lie in our power to achieve this analysis." Though of course the comparison to commensurable and incommensurable magnitudes erects a fresh

---

[1] Baumann, op. cit., Vol. II, p. 56 (Erdmann edn., p. 424).

[2] Baumann, op. cit., Vol. II, p. 57 (Erdmann edn., p. 83).

[3] Baumann, op. cit., Vol. II, p. 57 (Pertz edn., Vol. II, p. 55 [= Gerhardt edn., *phil. Schr.*, Vol. II, p. 62]).

* [*Identitäten*]

schreitbare Schranke zwischen zufälligen und nothwendigen Wahrheiten auf.

Sehr entschieden im Sinne der analytischen Natur der Zahlgesetze spricht sich W. Stanley Jevons aus*): „Zahl ist nur logische Unterscheidung und Algebra eine hoch entwickelte Logik."

§ 16. Aber auch diese Ansicht hat ihre Schwierigkeiten. Soll dieser hochragende, weitverzweigte und immer noch wachsende Baum der Zahlenwissenschaft in blossen Identitäten wurzeln? Und wie kommen die leeren Formen der Logik dazu, aus sich heraus solchen Inhalt zu gewinnen?

Mill meint: „Die Lehre, dass wir durch kunstfertiges Handhaben der Sprache Thatsachen entdecken, die verborgene Naturprocesse enthüllen können, ist dem gesunden Menschenverstande so entgegen, dass es schon einen Fortschritt in der Philosophie verlangt, um sie zu glauben."

Gewiss dann, wenn man sich bei dem kunstfertigen Handhaben nichts denkt. Mill wendet sich hier gegen einen Formalismus, der kaum von irgendwem vertreten wird. Jeder, der Worte oder mathematische Zeichen gebraucht, macht den Anspruch, dass sie etwas bedeuten, und niemand wird erwarten, dass aus leeren Zeichen etwas Sinnvolles hervorgehe. Aber es ist möglich, dass ein Mathematiker längere Rechnungen vollführt, ohne unter seinen Zeichen etwas sinnlich Wahrnehmbares, Anschauliches zu verstehen. Darum sind diese Zeichen noch nicht sinnlos; man unterscheidet dennoch ihren Inhalt von ihnen selbst, wenn dieser auch vielleicht nur mittels der Zeichen fassbar wird. Man ist sich bewusst, dass andere Zeichen für Dasselbe hätten festgesetzt werden können. Es genügt zu wissen, wie der in den Zeichen versinnlichte Inhalt logisch zu behandeln ist, und wenn man Anwendungen auf die Physik machen will, wie der Uebergang zu den Erscheinungen geschehen

*) The principles of science, London 1879, S. 156.

barrier between necessary and contingent truths, which for us at least is insuperable.

A very emphatic declaration in favour of the analytic nature of the laws of number is that of W. S. JEVONS[1]: "I hold that algebra is a highly developed logic, and number but logical discrimination."

§ 16. But this view, too, has its difficulties. Can the great tree of the science of number as we know it, towering, spreading, and still continually growing, have its roots in bare identities*? And how do the empty forms of logic come to disgorge so rich a content?

To quote MILL:[2] "The doctrine that we can discover facts, detect the hidden processes of nature, by an artful manipulation of language, is so contrary to common sense, that a person must have made some advances in philosophy to believe it."

Very true—if it be supposed that during the artful manipulation we do not think at all. MILL is here criticizing a kind of formalism that scarcely anyone would wish to defend. Everyone who uses words or mathematical symbols makes the claim that they mean something, and no one will expect any sense to emerge from empty symbols. But it is possible for a mathematician to perform quite lengthy calculations without understanding by his symbols anything intuitable, or with which we could be sensibly acquainted. And that does not mean that the symbols have no sense; we still distinguish between the symbols themselves and their content, even though it may be that the content can only be grasped by their aid. We realize perfectly that other symbols might have been assigned to stand for the same things. All we need to know is how to handle logically the content as made sensible in the symbols and, if we wish to apply our calculus to physics, how to effect the transition to the phenomena.

---

[1] *The Principles of Science*, London 1879, p. 156 [1874 edn., p. 174].
[2] Op. cit., Bk. II, cap. vi, § 2.

* [*Identitäten*]

muss. Aber in einer solchen Anwendung ist nicht der eigentliche Sinn der Sätze zu sehen. Dabei geht immer ein grosser Theil der Allgemeinheit verloren, und es kommt etwas Besonderes hinein, das bei andern Anwendungen durch Anderes ersetzt wird.

§ 17. Man kann trotz aller Herabsetzung der Deduction doch nicht leugnen, dass die durch Induction begründeten Gesetze nicht genügen. Aus ihnen müssen neue Sätze abgeleitet werden, die in keinem einzelnen von jenen enthalten sind. Dass sie in allen zusammen schon in gewisser Weise stecken, entbindet nicht von der Arbeit, sie daraus zu entwickeln und für sich herauszustellen. Damit eröffnet sich folgende Möglichkeit. Statt eine Schlussreihe unmittelbar an eine Thatsache anzuknüpfen, kann man, diese dahingestellt sein lassend, ihren Inhalt als Bedingung mitführen. Indem man so alle Thatsachen in einer Gedankenreihe durch Bedingungen ersetzt, wird man das Ergebniss in der Form erhalten, dass von einer Reihe von Bedingungen ein Erfolg abhängig gemacht ist. Diese Wahrheit wäre durch Denken allein, oder, um mit Mill zu reden, durch kunstfertiges Handhaben der Sprache begründet. Es ist nicht unmöglich, dass die Zahlgesetze von dieser Art sind. Sie wären dann analytische Urtheile, obwohl sie nicht durch Denken allein gefunden zu sein brauchten; denn nicht die Weise des Findens kommt hier in Betracht, sondern die Art der Beweisgründe; oder, wie Leibniz sagt*), „es handelt sich hier nicht um die Geschichte unserer Entdeckungen, die verschieden ist in verschiedenen Menschen, sondern um die Verknüpfung und die natürliche Ordnung der Wahrheiten, die immer dieselbe ist." Die Beobachtung hätte dann zuletzt zu entscheiden, ob die in dem so begründeten Gesetze enthaltenen Bedingungen erfüllt sind. So würde man schliesslich eben dahin gelangen, wohin man durch unmittelbare An-

*) Nouveaux Essais, IV, § 9; Erdm. S. 360.

It is, however, a mistake to see in such applications the real sense of the propositions; in any application a large part of their generality is always lost, and a particular element enters in, which in other applications is replaced by other particular elements.

§ 17. However much we may disparage deduction, it cannot be denied that the laws established by induction are not enough. New propositions must be derived from them which are not contained in any one of them by itself. No doubt these propositions are in a way contained covertly in the whole set taken together, but this does not absolve us from the labour of actually extracting them and setting them out in their own right. This seen, we can see also the following possibility. Instead of linking our chain of deductions direct to any matter of fact, we can leave the fact on one side, while adopting its content in the form of a condition. By substituting in this way conditions for facts throughout the whole of a train of reasoning, we shall finally reduce it to a form in which a certain result is made dependent on a certain series of conditions. This truth would be established by thought alone or, to use MILL's expression, by an artful manipulation of language. It is not impossible that the laws of number are of this type. This would make them analytic judgements, despite the fact that they would not normally be discovered by thought alone; for we are concerned here not with the way in which they are discovered but with the kind of ground on which their proof rests; or in LEIBNIZ's[1] words, "the question here is not one of the history of our discoveries, which is different in different men, but of the connexion and natural order of truths, which is always the same." It would then rest with observation finally to decide whether the conditions included in the laws thus established are actually fulfilled. Thus we should in the end arrive at the same position as we should have reached by linking our chain

[1] *Nouveaux Essais*, IV, § 9 (Erdmann edn., p. 362).

knüpfung der Schlussreihe an die beobachteten Thatsachen gekommen wäre. Aber die hier angedeutete Art des Vorgehens ist in vielen Fällen vorzuziehen, weil sie auf einen allgemeinen Satz führt, der nicht nur auf die grade vorliegenden Thatsachen anwendbar zu sein braucht. Die Wahrheiten der Arithmetik würden sich dann zu denen der Logik ähnlich verhalten wie die Lehrsätze zu den Axiomen der Geometrie. Jede würde in sich eine ganze Schlussreihe für den künftigen Gebrauch verdichtet enthalten, und ihr Nutzen würde darin bestehen, dass man die Schlüsse nicht mehr einzeln zu machen braucht, sondern gleich das Ergebniss der ganzen Reihe aussprechen kann*). Angesichts der gewaltigen Entwickelung der arithmetischen Lehren und ihrer vielfachen Anwendungen wird sich dann freilich die weit verbreitete Geringschätzung der analytischen Urtheile und das Märchen von der Unfruchtbarkeit der reinen Logik nicht halten lassen.

Wenn man diese nicht hier zuerst geäusserte Ansicht im Einzelnen so streng durchführen könnte, dass nicht der geringste Zweifel zurückbliebe, so würde das, wie mir scheint, kein ganz unwichtiges Ergebniss sein.

II. Meinungen einiger Schriftsteller über den Begriff der Anzahl.

§ 18. Indem wir uns nun den ursprünglichen Gegenständen der Arithmetik zuwenden, unterscheiden wir die einzelnen Zahlen 3, 4 u. s. f. von dem allgemeinen Begriffe

---

*) Es ist auffallend, dass auch Mill a. a. O. II. Buch, VI. Cap. § 4 diese Ansicht auszusprechen scheint. Sein gesunder Sinn durchbricht eben von Zeit zu Zeit sein Vorurtheil für das Empirische. Aber dieses bringt immer wieder Alles in Verwirrung, indem es ihn die physikalischen Anwendungen der Arithmetik mit dieser selbst verwechseln lässt. Er scheint nicht zu wissen, dass ein hypothetisches Urtheil auch dann wahr sein kann, wenn die Bedingung nicht wahr ist.

of deductions direct to observed matters of fact. But the type of procedure here indicated is in many cases to be preferred, because it leads to a general proposition, which need not be applicable only to the facts immediately under consideration. The truths of arithmetic would then be related to those of logic in much the same way as the theorems of geometry to the axioms. Each one would contain concentrated within it a whole series of deductions for future use, and the use of it would be that we need no longer make the deductions one by one, but can express simultaneously the result of the whole series.[1] If this be so, then indeed the prodigious development of arithmetical studies, with their multitudinous applications, will suffice to put an end to the widespread contempt for analytic judgements and to the legend of the sterility of pure logic.

This is not the first time that such a view has been put forward. If it could be worked out in detail, so rigorously that not the smallest doubt remained, that, it seems to me, would be a result not entirely without importance.

II. Views of certain writers on the concept of Number.

§ 18. On turning now to consider the primary objects of arithmetic, we must distinguish between the individual numbers 3, 4 and so on, and the general concept of Number.

---

[1] It is remarkable that MILL too (op. cit., Bk. II, cap. vi, § 4) seems to express this view. His sound sense, in fact, from time to time breaks through his prejudice in favour of the empirical. But this same prejudice as often brings everything back into a muddle, by making him confuse the physical applications of arithmetic with arithmetic itself. He seems to be unaware that a hypothetical judgement can be true even when the antecedent is not true.

der Anzahl. Nun haben wir uns schon dafür entschieden, dass die einzelnen Zahlen am besten in der Weise von Leibniz, Mill, H. Grassmann und Andern aus der Eins und der Vermehrung um eins abgeleitet werden, dass aber diese Erklärungen unvollständig bleiben, solange die Eins und die Vermehrung um eins unerklärt sind. Wir haben gesehen, dass man allgemeiner Sätze bedarf, um aus diesen Definitionen die Zahlformeln abzuleiten. Solche Gesetze können eben wegen ihrer Allgemeinheit nicht aus den Definitionen der einzelnen Zahlen folgen, sondern nur aus dem allgemeinen Begriffe der Anzahl. Wir unterwerfen diesen jetzt einer genaueren Betrachtung. Dabei werden voraussichtlich auch die Eins und die Vermehrung um eins erörtert werden müssen und somit auch die Definitionen der einzelnen Zahlen eine Ergänzung zu erwarten haben.

§ 19. Hier möchte ich mich nun gleich gegen den Versuch wenden, die Zahl geometrisch als Verhältnisszahl von Längen oder Flächen zu fassen. Man glaubte offenbar die vielfachen Anwendungen der Arithmetik auf Geometrie dadurch zu erleichtern, dass man gleich die Anfänge in die engste Beziehung setzte.

Newton*) will unter Zahl nicht so sehr eine Menge von Einheiten als das abstracte Verhältniss einer jeden Grösse zu einer andern derselben Art verstehen, die als Einheit genommen wird. Man kann zugeben, dass hiermit die Zahl im weitern Sinne, wozu auch die Brüche und Irrationalzahlen gehören, zutreffend beschrieben sei; doch werden hierbei die Begriffe der Grösse und des Grössenverhältnisses vorausgesetzt. Danach scheint es, dass die Erklärung der Zahl im engern Sinne, der Anzahl, nicht überflüssig werde; denn Euklid braucht den Begriff des Gleichvielfachen um die Gleichheit von zwei Längenverhältnissen zu definiren; und das Gleichvielfache kommt

*) Baumann a. a. O. Bd. I, S. 475.

Now we have already decided in favour of the view that the individual numbers are best derived, in the way proposed by LEIBNIZ, MILL, H. GRASSMANN and others, from the number one together with increase by one, but that these definitions remain incomplete so long as the number one and increase by one are themselves undefined. And we have seen that we have need of general propositions if we are to derive the numerical formulae from these definitions. Such laws cannot, just because of their generality, follow from the definitions of the individual numbers, but only from the general concept of Number. It is this concept that we shall now submit to a closer examination; in the course of this we may expect to have also to discuss the number one and increase by one, as a result of which in turn we shall expect to complete the definitions of the individual numbers.

§ 19. At this point I should like straight away to oppose the attempt to think of number geometrically, as a ratio between lengths or surfaces. Obviously, the thought behind this was to facilitate the numerous applications of arithmetic to geometry by putting the rudiments of both in the closest connexion from the outset.

NEWTON[1] proposes to understand by number not so much a set of units as the relation in the abstract between any given magnitude* and another magnitude of the same kind which is taken as unity. It may be granted that this is an apt description of number in the wider sense, in which it includes [besides the integers] also fractions and irrational numbers; but it presupposes the concepts of magnitude and of relation in respect of magnitude. This should presumably mean that a definition of number in the narrower sense, or cardinal Number, will still be needed; for EUCLID**, in order to define the identity of two ratios between lengths, makes use of the concept of equimultiples, and equimultiples bring us back once again

---

[1] Baumann, op. cit., Vol. I, p. 475 [*Arithmetica Universalis*, Vol. I, cap. ii, 3.]

* [*quantitas*]

** [*Elements*, Bk. V., Def. 5.]

wieder auf eine Zahlengleichheit hinaus. Aber es mag sein, dass die Gleichheit von Längenverhältnissen unabhängig vom Zahlbegriffe definirbar ist. Man bliebe dann jedoch im Ungewissen darüber, in welcher Beziehung die so geometrisch definirte Zahl zu der Zahl des gemeinen Lebens stände. Dies wäre dann ganz von der Wissenschaft getrennt. Und doch kann man wohl von der Arithmetik verlangen, dass sie die Anknüpfungspunkte für jede Anwendung der Zahl bieten muss, wenn auch die Anwendung selbst nicht ihre Sache ist. Auch das gewöhnliche Rechnen muss die Begründung seines Verfahrens in der Wissenschaft finden. Und dann erhebt sich die Frage, ob die Arithmetik selbst mit einem geometrischen Begriffe der Zahl auskomme, wenn man an die Anzahl der Wurzeln einer Gleichung, der Zahlen, die prim zu einer Zahl und kleiner als sie sind, und ähnliche Vorkommnisse denkt. Dagegen kann die Zahl, welche die Antwort auf die Frage wieviel? giebt, auch bestimmen, wieviel Einheiten in einer Länge enthalten sind. Die Rechnung mit negativen, gebrochenen, Irrationalzahlen kann auf die mit den natürlichen Zahlen zurückgeführt werden. Newton wollte aber vielleicht unter Grössen, als deren Verhältniss die Zahl definirt wird, nicht nur geometrische, sondern auch Mengen verstehen. Dann wird jedoch die Erklärung für unsern Zweck unbrauchbar, weil von den Ausdrücken „Zahl durch die eine Menge bestimmt wird" und „Verhältniss einer Menge zur Mengeneinheit" der letztere keine bessere Auskunft als der erstere giebt.

§ 20. Die erste Frage wird nun sein, ob Zahl definirbar ist. Hankel*) spricht sich dagegen aus: „Was es heisst, ein Object 1 mal, 2 mal, 3 mal . . . denken oder setzen, kann bei der principiellen Einfachheit des Begriffes der Setzung nicht definirt werden." Hier kommt es jedoch weniger auf das Setzen als auf das 1 mal, 2 mal, 3 mal an. Wenn dies

* Theorie der complexen Zahlensysteme, S. 1.

to numerical identity. However, let it be, as it may be, the case that identity of ratios between lengths can in fact be defined without any reference to the concept of number. Even so, we should still remain in doubt as to how the number defined geometrically in this way is related to the number of ordinary life, which would then be entirely cut off from science. Yet surely we are entitled to demand of arithmetic that its numbers should be adapted for use in every application made of number, even although that application is not itself the business of arithmetic. Even in our everyday sums, we must be able to rely on the science of arithmetic to provide the basis for the methods we use. And moreover, the question arises whether arithmetic itself can make do with a geometrical concept of number, when we think of some of the notions that occur in it, such as the Number of roots of an equation or of numbers prime to and smaller than a given number. On the other hand, the number which gives the answer to the question *How many?* can answer among other things how many units are contained in a length. And operations with negative, fractional and irrational numbers can all be reduced to operations with the natural numbers. Perhaps what NEWTON wished to understand by magnitudes, in defining number as a relation between magnitudes, was not geometrical magnitudes only, but also sets. In that case, however, his definition is useless for our purposes, since the expression "relation between a set and the unit of the set" tells us no more than the expression "number by which a set is determined."

§ 20. The first question to be faced, then, is whether number is definable. HANKEL[1] declares that it is not, in these words: "What we mean by thinking or putting a thing once, twice, three times, and so on, cannot be defined, because of the simplicity in principle of the concept of putting." But the point is surely not so much the putting as the once, twice and three times. If this could be defined, the indefinability

[1] Op. cit., p. 1.

definirt werden könnte, so würde die Undefinirbarkeit des Setzens uns wenig beunruhigen. Leibniz ist geneigt, die Zahl wenigstens annähernd als adaequate Idee anzusehen, d. h. als eine solche, die so deutlich ist, dass alles, was in ihr vorkommt, wieder deutlich ist.

Wenn man im Ganzen mehr dazu neigt, die Anzahl für undefinirbar zu halten, so liegt das wohl mehr an dem Misslingen darauf gerichteter Versuche als an dem Bestehen der Sache selbst entnommener Gegengründe.

Ist die Anzahl eine Eigenschaft der äusseren Dinge?

§ 21. Versuchen wir wenigstens der Anzahl ihre Stelle unter unsern Begriffen anzuweisen! In der Sprache erscheinen Zahlen meistens in adjectivischer Form und in attributiver Verbindung ähnlich wie die Wörter hart, schwer, roth, welche Eigenschaften der äusseren Dinge bedeuten. Es liegt die Frage nahe, ob man die einzelnen Zahlen auch so auffassen müsse, und ob demgemäss der Begriff der Anzahl etwa mit dem der Farbe zusammengestellt werden könne.

Dies scheint die Meinung von M. Cantor*) zu sein, wenn er die Mathematik eine Erfahrungswissenschaft nennt, insofern sie von der Betrachtung von Objecten der Aussenwelt ihren Anfang nehme. Nur durch Abstraction von Gegenständen entstehe die Zahl.

E. Schröder**) lässt die Zahl der Wirklichkeit nachgebildet, aus ihr entnommen werden, indem die Einheiten durch Einer abgebildet würden. Dies nennt er Abstrahiren der Zahl. Bei dieser Abbildung würden die Einheiten nur in Hinsicht ihrer Häufigkeit dargestellt, indem von allen

---

*) Grundzüge einer Elementarmathematik, S. 2, § 4. Aehnlich Lipschitz, Lehrbuch der Analysis, Bonn 1877, S. 1.

**) Lehrbuch der Arithmetik und Algebra, Leipz. 1873, S. 6, 10 u. 11.

of putting would scarcely worry us. LEIBNIZ is inclined to regard number as an adequate idea, meaning one which is so clear that every element contained in it is also clear, or at least as an almost adequate one.

If the general inclination is, on the whole, to hold that Number is indefinable, that is more because attempts to define it have failed than because anything has been discovered in the nature of the case to show that it must be so.

Is Number a property of external things?

§ 21. Let us try at least to assign to Number its proper place among our concepts. In language, numbers most commonly appear in adjectival form and attributive construction in the same sort of way as the words hard or heavy or red, which have for their meanings properties of external things. It is natural to ask whether we must think of the individual numbers too as such properties, and whether, accordingly, the concept of Number can be classed along with that, say, of colour.

That it can, seems to be the view of M. CANTOR,[1] when he calls mathematics an empirical science in so far as it begins with the consideration of things in the external world. On his view, number originates only by abstraction from objects.

For E. SCHRÖDER[2] number is modelled on actuality, derived from it by a process of copying the actual units with ones, which he calls the abstraction of number. In this copying, the units are only represented in point of their frequency, all

---

[1] *Grundzüge einer Elementarmathematik*, p. 2, § 4. Similarly Lipschitz, op. cit., p. 1.

[2] Op. cit., pp. 6, 10–11.

andern Bestimmungen der Dinge als Farbe, Gestalt abgesehen werde. Hier ist Häufigkeit nur ein anderer Ausdruck für Anzahl. Schröder stellt also Häufigkeit oder Anzahl in eine Linie mit Farbe und Gestalt und betrachtet sie als eine Eigenschaft der Dinge.

§ 22. Baumann*) verwirft den Gedanken, dass die Zahlen von den äussern Dingen abgezogene Begriffe seien: „Weil nämlich die äussern Dinge uns keine strengen Einheiten darstellen; sie stellen uns abgegränzte Gruppen oder sinnliche Punkte dar, aber wir haben die Freiheit, diese selber wieder als Vieles zu betrachten." In der That, während ich nicht im Stande bin, durch blosse Auffassungsweise die Farbe eines Dinges oder seine Härte im Geringsten zu verändern, kann ich die Ilias als Ein Gedicht, als 24 Gesänge oder als eine grosse Anzahl von Versen auffassen. Spricht man nicht in einem ganz andern Sinne von 1000 Blättern als von grünen Blättern des Baumes? Die grüne Farbe legen wir jedem Blatte bei, nicht so die Zahl 1000. Wir können alle Blätter des Baumes unter dem Namen seines Laubes zusammenfassen. Auch dieses ist grün, aber nicht 1000. Wem kommt nun eigentlich die Eigenschaft 1000 zu? Fast scheint es weder dem einzelnen Blatte noch der Gesammtheit; vielleicht gar nicht eigentlich den Dingen der Aussenwelt? Wenn ich jemandem einen Stein gebe mit den Worten: bestimme das Gewicht hiervon, so habe ich ihm damit den ganzen Gegenstand seiner Untersuchung gegeben. Wenn ich ihm aber einen Pack Spielkarten in die Hand gebe mit den Worten: bestimme die Anzahl hiervon, so weiss er nicht, ob ich die Zahl der Karten oder der vollständigen Spiele oder etwa der Wertheinheiten beim Skatspiele erfahren will. Damit, dass ich ihm den Pack in die Hand gebe, habe ich ihm den Gegenstand seiner Untersuchung noch nicht vollständig gegeben; ich muss ein Wort:

*) A. a. O. Bd. II, S. 669.

other properties of the things concerned, such as their colour or shape, being disregarded. Here frequency is only another name for Number. It follows, therefore, that SCHRÖDER puts frequency or Number on a level with colour and shape, and treats it as a property of things.

§ 22. BAUMANN[1] rejects the view that numbers are concepts extracted from external things: "The reason being that external things do not present us with any strict units; they present us with isolated groups or sensible points, but we are at liberty to treat each one of these itself again as a many." And it is quite true that, while I am not in a position, simply by thinking of it differently, to alter the colour or hardness of a thing in the slightest, I am able to think of the Iliad either as one poem, or as 24 Books, or as some large Number of verses. Is it not in totally different senses that we speak of a tree as having 1000 leaves and again as having green leaves? The green colour we ascribe to each single leaf, but not the number 1000. If we call all the leaves of a tree taken together its foliage, then the foliage too is green, but it is not 1000. To what then does the property 1000 really belong? It almost looks as though it belongs neither to any single one of the leaves nor to the totality of them all; is it possible that it does not really belong to things in the external world at all? If I give someone a stone with the words: Find the weight of this, I have given him precisely the object he is to investigate. But if I place a pile of playing cards in his hands with the words: Find the Number of these, this does not tell him whether I wish to know the number of cards, or of complete packs of cards, or even say of points in the game of skat. To have given him the pile in his hands is not yet to have given him completely the object he is to investigate; I must add some

---

[1] Op. cit., Vol. II, p. 669.

Karte, Spiel, Wertheinheit hinzufügen. Man kann auch nicht sagen, dass die verschiedenen Zahlen hier so wie die verschiedenen Farben neben einander bestehen. Auf die einzelne farbige Fläche kann ich hindeuten, ohne ein Wort zu sagen, nicht so auf die einzelne Zahl. Wenn ich einen Gegenstand mit demselben Rechte grün und roth nennen kann, so ist das ein Zeichen, dass dieser Gegenstand nicht der eigentliche Träger des Grünen ist. Diesen habe ich erst in einer Fläche, die nur grün ist. So ist auch ein Gegenstand, dem ich mit demselben Rechte verschiedene Zahlen zuschreiben kann, nicht der eigentliche Träger einer Zahl.

Ein wesentlicher Unterschied zwischen Farbe und Anzahl besteht demnach darin, dass die blaue Farbe einer Fläche unabhängig von unserer Willkühr zukommt. Sie ist ein Vermögen, gewisse Lichtstrahlen zurückzuwerfen, andere mehr oder weniger zu verschlucken, und daran kann unsere Auffassung nicht das Geringste ändern. Dagegen kann ich nicht sagen, dass dem Pack Spielkarten die Anzahl 1 oder 100 oder irgend eine andere an sich zukomme, sondern höchstens in Bezug auf unsere willkührliche Auffassungsweise, und dann auch nicht so, dass wir ihm die Anzahl einfach als Praedicat beilegen könnten. Was wir ein vollständiges Spiel nennen wollen, ist offenbar eine willkührliche Festsetzung und der Pack Spielkarten weiss nichts davon. Indem wir ihn aber in dieser Hinsicht betrachten, entdecken wir vielleicht, dass wir ihn zwei vollständige Spiele nennen können. Jemand, der nicht wüsste, was man ein vollständiges Spiel nennt, würde wahrscheinlich irgend eine andere Anzahl eher an ihm herausfinden, als grade die Zwei.

§ 23. Die Frage, wem die Zahl als Eigenschaft zukomme, beantwortet Mill*) so:

„Der Name einer Zahl bezeichnet eine Eigenschaft, die dem Aggregat von Dingen angehört, welche wir mit

*) A. a. O. III. Buch, XXIV. Cap., § 5.

further word—cards, or packs, or points. Nor can we say that in this case the different numbers exist in the same thing side by side, as different colours do. I can point to the patch of each individual colour without saying a word, but I cannot in the same way point to the individual numbers. If I can call the same object red and green with equal right, it is a sure sign that the object named is not what really has the green colour; for that we must first get a surface which is green only. Similarly, an object to which I can ascribe different numbers with equal right is not what really has a number.

It marks, therefore, an important difference between colour and Number, that a colour such as blue belongs to a surface independently of any choice of ours. The blue colour is a power of reflecting light of certain wavelengths and of absorbing to varying extents light of other wavelengths; to this, our way of regarding it cannot make the slightest difference. The Number 1, on the other hand, or 100 or any other Number, cannot be said to belong to the pile of playing cards in its own right, but at most to belong to it in view of the way in which we have chosen to regard it; and even then not in such a way that we can simply assign the Number to it as a predicate. What we choose to call a complete pack is obviously an arbitrary decision, in which the pile of playing cards has no say. But it is when we examine the pile in the light of this decision, that we discover perhaps that we can call it two complete packs. Anyone who did not know what we call a complete pack would probably discover in the pile any other Number you like before hitting on two.

§ 23. To the question: What is it that the number belongs to as a property? MILL[1] replies as follows: the name of a number connotes, "of course, some property belonging to the

---

[1] Op. cit., Bk. III, cap. xxiv, § 5.

dem Namen benennen; und diese Eigenschaft ist die charakteristische Weise, in welcher das Aggregat zusammengesetzt ist oder in Theile zerlegt werden kann."

Hier ist zunächst der bestimmte Artikel in dem Ausdrucke „die charakteristische Weise" ein Fehler; denn es giebt sehr verschiedene Weisen, wie man ein Aggregat zerlegen kann, und man kann nicht sagen, dass Eine allein charakteristisch wäre. Ein Bündel Stroh kann z. B. so zerlegt werden, dass man alle Halme durchschneidet, oder so, dass man es in einzelne Halme auflöst, oder so dass man zwei Bündel daraus macht. Ist denn ein Haufe von hundert Sandkörnern ebenso zusammengesetzt wie ein Bündel von 100 Strohhalmen? und doch hat man dieselbe Zahl. Das Zahlwort „Ein" in dem Ausdruck „Ein Strohhalm" drückt doch nicht aus, wie dieser Halm aus Zellen oder aus Molekeln zusammengesetzt ist. Noch mehr Schwierigkeit macht die Zahl 0. Müssen denn die Strohhalme überhaupt ein Bündel bilden, um gezählt werden zu können? Muss man die Blinden im Deutschen Reiche durchaus in einer Versammlung vereinigen, damit der Ausdruck „Zahl der Blinden im Deutschen Reiche" einen Sinn habe? Sind tausend Weizenkörner, nachdem sie ausgesäet sind, nicht mehr tausend Weizenkörner? Giebt es eigentlich Aggregate von Beweisen eines Lehrsatzes oder von Ereignissen? und doch kann man auch diese zählen. Dabei ist es gleichgiltig, ob die Ereignisse gleichzeitig oder durch Jahrtausende getrennt sind.

§ 24. Damit kommen wir auf einen andern Grund, die Zahl nicht mit Farbe und Festigkeit zusammenzustellen: die bei weitem grössere Anwendbarkeit.

Mill*) meint, die Wahrheit, dass, was aus Theilen zusammengesetzt ist, aus Theilen dieser Theile zusammengesetzt ist, sei von allen Naturerscheinungen giltig, weil

*) A. a. O. III. Buch, XXIV. Cap. § 5.

agglomeration of things which we call by the name; and that property is the characteristic manner in which the agglomeration is made up of, and may be separated into, parts."

Here the definite article in the phrase "the characteristic manner" is a mistake right away; for there are very various manners in which an agglomeration can be separated into parts, and we cannot say that one alone would be characteristic. For example, a bundle of straw can be separated into parts by cutting all the straws in half, or by splitting it up into single straws, or by dividing it into two bundles. Further, is a heap of a hundred grains of sand made up of parts in exactly the same way as a bundle of 100 straws? And yet we have the same number. The number word "one", again, in the expression "one straw" signally fails to do justice to the way in which the straw is made up of cells or molecules. Still more difficulty is presented by the number o. Besides, need the straws form any sort of bundle at all in order to be numbered? Must we literally hold a rally of all the blind in Germany before we can attach any sense to the expression "the number of blind in Germany"? Are a thousand grains of wheat, when once they have been scattered by the sower, a thousand grains of wheat no longer? Do such things really exist as agglomerations of proofs of a theorem, or agglomerations of events? And yet these too can be numbered. Nor does it make any difference whether the events occur together or thousands of years apart.

§ 24. This brings us to another reason for refusing to class number along with colour and solidity: it is applicable over a far wider range.

MILL[1] maintains that the truth that whatever is made up of parts is made up of parts of those parts holds good for natural phenomena of every sort, since all admit of being

---

[1] Op. cit., Bk. III, cap. xxiv, § 5.

alle gezählt werden könnten. Aber kann nicht noch weit mehr gezählt werden? Locke*) sagt: „Die Zahl findet Anwendung auf Menschen, Engel, Handlungen, Gedanken, jedes Ding, das existirt oder vorgestellt werden kann". Leibniz**) verwirft die Meinung der Scholastiker, dass die Zahl auf unkörperliche Dinge unanwendbar sei, und nennt die Zahl gewissermaassen eine unkörperliche Figur, entstanden aus der Vereinigung irgendwelcher Dinge, z. B. Gottes, eines Engels, eines Menschen, der Bewegung, welche zusammen vier sind. Daher, meint er, ist die Zahl etwas ganz Allgemeines und zur Metaphysik gehörig. An einer andern Stelle***) sagt er: „Gewogen kann nicht werden, was nicht Kraft und Vermögen hat; was keine Theile hat, hat demgemäss kein Maass; aber es giebt nichts, was nicht die Zahl zulässt. So ist die Zahl gleichsam die metaphysische Figur".

Es wäre in der That wunderbar, wenn eine, von äussern Dingen abstrahirte Eigenschaft, auf Ereignisse, auf Vorstellungen, auf Begriffe ohne Aenderung des Sinnes übertragen werden könnte. Es wäre grade so, also ob man von einem schmelzbaren Ereignisse, einer blauen Vorstellung, einem salzigen Begriffe, einem zähen Urtheile reden wollte.

Es ist ungereimt, dass an Unsinnlichem vorkomme, was seiner Natur nach sinnlich ist. Wenn wir eine blaue Fläche sehen, so haben wir einen eigenthümlichen Eindruck, der dem Worte „blau" entspricht; und diesen erkennen wir wieder, wenn wir eine andere blaue Fläche erblicken. Wollten wir annehmen, dass in derselben Weise beim Anblick eines Dreiecks etwas Sinnliches dem Worte „drei" entspräche, so müssten wir dies auch in drei Begriffen wiederfinden; etwas Unsinnliches würde etwas Sinnliches an sich haben.

*) Baumann a. a. O. Bd. I, S. 409.
**) Ebenda, Bd. II, S. 56.
***) Ebenda, Bd. II, S. 2.

numbered. But cannot still far more than this be numbered? LOCKE[1] says: "Number applies itself to men, angels, actions, thoughts—everything that either doth exist or can be imagined." LEIBNIZ[2] rejects the view of the schoolmen that number is not applicable to immaterial things, and calls number a sort of immaterial figure, which results from the union of things of any sorts whatsoever, for example of God, an angel, a man and motion, which together are four. For which reason he holds that number is of supreme universality and belongs to metaphysics. In another passage[3] he says: "Some things cannot be weighed, as having no force and power; some things cannot be measured, by reason of having no parts; but there is nothing which cannot be numbered. Thus number is, as it were, a kind of metaphysical figure."

It would indeed be remarkable if a property abstracted from external things could be transferred without any change of sense to events, to ideas and to concepts. The effect would be just like speaking of fusible events, or blue ideas, or salty concepts or tough judgements.

It does not make sense that what is by nature sensible should occur in what is non-sensible. When we see a blue surface, we have an impression of a unique sort, which corresponds to the word "blue"; this impression we recognize again, when we catch sight of another blue surface. In order to suppose that there is in the same way, when we look at a triangle, something sensible corresponding to the word "three", we should have to commit ourselves to finding that same thing again in three concepts too; so that something non-sensible would have in it something sensible. It may

[1] Baumann, op. cit., Vol. 1, p. 409. [*Essay*, Bk. II, cap. xvi, § 1].
[2] Baumann, op. cit., Vol. II, pp. 2–3 [Erdmann edn., p. 8].
[3] Baumann, op. cit., Vol. II, p. 56 [Erdmann edn., p. 162].

Man kann wohl zugeben, dass dem Worte „dreieckig" eine Art sinnlicher Eindrücke entspreche, aber man muss dabei dies Wort als Ganzes nehmen. Die Drei darin sehen wir nicht unmittelbar; sondern wir sehen etwas, woran eine geistige Thätigkeit anknüpfen kann, welche zu einem Urtheile führt, in dem die Zahl 3 vorkommt. Womit nehmen wir denn etwa die Anzahl der Schlussfiguren wahr, die Aristoteles aufstellt? etwa mit den Augen? wir sehen höchstens gewisse Zeichen für diese Schlussfiguren, nicht sie selbst. Wie sollen wir ihre Anzahl sehen können, wenn sie selbst unsichtbar bleiben? Aber, meint man vielleicht, es genügt, die Zeichen zu sehen; deren Zahl ist gleich der Zahl der Schlussfiguren. Woher weiss man denn das? Dazu muss man doch schon auf andere Weise die letztere bestimmt haben. Oder ist der Satz „die Anzahl der Schlussfiguren ist vier" nur ein anderer Ausdruck für „die Anzahl der Zeichen der Schlussfiguren ist vier"? Nein! von den Zeichen soll nichts ausgesagt werden; von den Zeichen will niemand etwas wissen, wenn nicht deren Eigenschaft zugleich eine des Bezeichneten ausdrückt. Da ohne logischen Fehler dasselbe verschiedene Zeichen haben kann, braucht nicht einmal die Zahl der Zeichen mit der des Bezeichneten übereinzustimmen.

§ 25. Während für Mill die Zahl etwas Physikalisches ist, besteht sie für Locke und Leibniz nur in der Idee. In der That sind, wie Mill*) sagt, zwei Aepfel von drei Aepfeln, zwei Pferde von einem Pferd physikalisch verschieden, ein davon verschiedenes sichtliches und fühlbares Phänomen**). Aber ist daraus zu schliessen, dass die

*) A. a. O. III. Buch, XXIV. Cap. § 5.

**) Genau genommen müsste hinzugefügt werden: sobald sie überhaupt ein Phänomen sind. Wenn aber Jemand ein Pferd in Deutschland und eines in Amerika (und sonst keins) hat, so besitzt er zwei Pferde. Diese bilden jedoch kein Phänomen, sondern nur jedes Pferd für sich könnte so gennant werden.

certainly be granted that a sensible impression of a sort does correspond to the word "triangular", but then the word must be taken as a whole. The three in it we do not see directly; rather, we see something upon which can fasten an intellectual activity of ours leading to a judgement in which the number 3 occurs. How is it after all that we do become acquainted with, let us say, the Number of figures of the syllogism as drawn up by Aristotle? Is it perhaps with our eyes? What we see is at most certain symbols for the syllogistic figures, not the figures themselves. How are we to be able to see their Number, if they themselves remain invisible? However, it may be argued that it is enough to see the symbols; their number is identical with the number of the figures. But then, how do we know this? For that, we must have already ascertained the number of the figures by some other means. Or is the proposition "The Number of figures of the syllogism is four" only another way of putting the proposition that "The Number of symbols for figures of the syllogism is four?" Of course it is not. There is no intention of saying anything about the symbols; no one wants to know anything about them, except in so far as some property of theirs directly mirrors some property in what they symbolize. Besides, the same thing can, without any logical fallacy, be symbolized by several different symbols, so that there is not even any need for the number of symbols to coincide with the number of things symbolised.

§ 25. While for MILL the number is something physical, for LOCKE and LEIBNIZ it exists only as a notion.* MILL[1] is, of course, quite right that two apples are physically different from three apples, and two horses from one horse; that they are a different visible and tangible phenomenon.[2] But are

---

[1] Op. cit., Bk. III, cap. xxiv, § 5.

[2] Strictly speaking we should add: provided that they are a phenomenon at all. For if someone has one horse in Germany and one in America (and no others), then he does possess two horses; yet these two horses do not form a phenomenon,—only each one of the two by itself could be so described.

* [*in der Idee*]

Zweiheit, Dreiheit, etwas Physikalisches ist? Ein Paar Stiefel kann dieselbe sichtbare und fühlbare Erscheinung sein, wie zwei Stiefel. Hier haben wir einen Zahlenunterschied, dem kein physikalischer entspricht; denn zwei und Ein Paar sind keineswegs dasselbe, wie Mill sonderbarer Weise zu glauben scheint. Wie ist es endlich möglich, dass sich zwei Begriffe von drei Begriffen physikalisch unterscheiden?

So sagt Berkeley*): „Es ist zu bemerken, dass die Zahl nichts Fixes und Festgestelltes ist, was realiter in den Dingen selber existirte. Sie ist gänzlich Geschöpf des Geistes, wenn er entweder eine Idee an sich oder eine Combination von Ideen betrachtet, der er einen Namen geben will und sie so für eine Einheit gelten lässt. Jenachdem der Geist seine Ideen variirend combinirt, variirt die Einheit, und wie die Einheit so variirt auch die Zahl, welche nur eine Sammlung von Einheiten ist. Ein Fenster = 1; ein Haus, in dem viele Fenster sind, = 1; viele Häuser machen Eine Stadt aus."

Ist die Zahl etwas Subjectives?

§ 26. In diesem Gedankengange kommt man leicht dazu, die Zahl für etwas Subjectives anzusehen. Es scheint die Weise, wie die Zahl in uns entsteht, über ihr Wesen Aufschluss geben zu können. Auf eine psychologische Untersuchung also würde es dann ankommen. In diesem Sinne sagt wohl Lipschitz**):

„Wer über gewisse Dinge einen Ueberblick gewinnen will, der wird mit einem bestimmten Dinge beginnen und immer ein neues Ding den früheren hinzufügen". Dies scheint viel besser darauf zu passen, wie wir etwa die Anschauung eines Sternbildes erhalten, als auf die Zahlbildung. Die

---

*) Baumann a. a. O. Bd. II. S. 428.
**) Lehrbuch der Analysis, S. 1. Ich nehme an, dass Lipschitz einen innern Vorgang im Sinne hat.

we to infer from this that their twoness or threeness is something physical? *One* pair of boots may be the same visible and tangible phenomenon as *two* boots. Here we have a difference in number to which no physical difference corresponds; for *two* and *one pair* are by no means the same thing, as MILL seems oddly to believe. How is it possible, after all, for two concepts to be physically distinguishable from three concepts?

To quote BERKELEY[1]: "It ought to be considered that number ... is nothing fixed and settled, really existing in things themselves. It is entirely the creature of the mind, considering, either an idea by itself, or any combination of ideas to which it gives one name, and so makes it pass for a unit. According as the mind variously combines its ideas, the unit varies; and as the unit, so the number, which is only a collection of units, doth also vary. We call a window one, a chimney one, and yet a house in which there are many windows, and many chimneys, hath an equal right to be called one, and many houses go to the making of one city."

Is number something subjective?

§ 26. This line of thought may easily lead us to regard number as something subjective. It looks as though the way in which number originates in us may prove the key to its essential nature. The matter would thus become one for a psychological enquiry. This is indeed what LIPSCHITZ[2] is thinking of when he writes: "Anyone who proposes to make a survey of a number of things, will begin with some one particular thing and proceed by continually adding a new one to those previously selected." This seems to describe much better how we acquire say the intuition of a constellation than how we construct numbers. The intention to make a

---

[1] Baumann, op. cit., Vol. II, p. 428 [*New Theory of Vision*, § 109].

[2] Op. cit., p. 1. I take it that Lipschitz means to refer to a mental process.

Absicht, einen Ueberblick zu gewinnen, ist unwesentlich; denn man wird kaum sagen können, dass eine Herde übersichtlicher wird, wenn man erfährt, aus wieviel Häuptern sie besteht.

Eine solche Beschreibung der innern Vorgänge, die der Fällung eines Zahlurtheils vorhergehen, kann nie, auch wenn sie zutreffender ist, eine eigentliche Begriffsbestimmung ersetzen. Sie wird nie zum Beweise eines arithmetischen Satzes herangezogen werden können; wir erfahren durch sie keine Eigenschaft der Zahlen. Denn die Zahl ist so wenig ein Gegenstand der Psychologie oder ein Ergebniss psychischer Vorgänge, wie es etwa die Nordsee ist. Der Objectivität der Nordsee thut es keinen Eintrag, dass es von unserer Willkühr abhangt, welchen Theil der allgemeinen Wasserbedeckung der Erde wir abgrenzen und mit dem Namen „Nordsee" belegen wollen. Das ist kein Grund, dies Meer auf psychologischem Wege erforschen zu wollen. So ist auch die Zahl etwas Objectives. Wenn man sagt „die Nordsee ist 10,000 Quadratmeilen gross," so deutet man weder durch „Nordsee" noch durch „10,000" auf einen Zustand oder Vorgang in seinem Innern hin, sondern man behauptet etwas ganz Objectives, was von unsern Vorstellungen und dgl. unabhängig ist. Wenn wir etwa ein ander Mal die Grenzen der Nordsee etwas anders ziehen oder unter „10,000" etwas Anderes verstehen wollten, so würde nicht derselbe Inhalt falsch, der vorher richtig war; sondern an die Stelle eines wahren Inhalts wäre vielleicht ein falscher geschoben, wodurch die Wahrheit jenes ersteren in keiner Weise aufgehoben würde.

Der Botaniker will etwas ebenso Thatsächliches sagen, wenn er die Anzahl der Blumenblätter einer Blume, wie wenn er ihre Farbe angiebt. Das eine hangt so wenig wie das andere von unserer Willkühr ab. Eine gewisse Aehnlichkeit der Anzahl und der Farbe ist also da; aber diese besteht nicht darin, dass beide an äusseren Dingen sinnlich wahrnehmbar, sondern darin, dass beide objectiv sind.

survey is not essential; for it will scarcely be maintained that it becomes any easier to survey a herd after we have learned how many head it comprises.

No description of this kind of the mental processes which precede the forming of a judgement of number*, even if more to the point than this one, can ever take the place of a genuine definition of the concept. It can never be adduced in proof of any proposition of arithmetic; it acquaints us with none of the properties of numbers. For number is no whit more an object of psychology or a product of mental processes than, let us say, the North Sea is. The objectivity of the North Sea is not affected by the fact that it is a matter of our arbitrary choice which part of all the water on the earth's surface we mark off and elect to call the "North Sea". This is no reason for deciding to investigate the North Sea by psychological methods. In the same way number, too, is something objective. If we say "The North Sea is 10,000 square miles in extent" then neither by "North Sea" nor by "10,000" do we refer to any state of or process in our minds: on the contrary, we assert something quite objective, which is independent of our ideas and everything of the sort. If we should happen to wish, on another occasion, to draw the boundaries of the North Sea differently or to understand something different by "10,000", that would not make false the same content that was previously true: what we should perhaps rather say is, that a false content had now taken the place of a true, without in any way depriving its predecessor of its truth.

The botanist means to assert something just as factual when he gives the Number of a flower's petals as when he gives their colour. The one depends on our arbitrary choice just as little as the other. There does, therefore, exist a certain similarity between Number and colour; it consists, however, not in our becoming acquainted with them both in external things through the senses, but in their being both objective.

* [For Frege, a "judgement of number" (*Zahlurtheil*), like its verbal expression, a "statement of number" (*Zahlangabe*), is one as to *how many* of something there are.]

Ich unterscheide das Objective von dem Handgreiflichen, Räumlichen, Wirklichen. Die Erdaxe, der Massenmittelpunkt des Sonnensystems sind objectiv, aber ich möchte sie nicht wirklich nennen, wie die Erde selbst. Man nennt den Aequator oft eine gedachte Linie; aber es wäre falsch, ihn eine erdachte Linie zu nennen; er ist nicht durch Denken entstanden, das Ergebniss eines seelischen Vorgangs, sondern nur durch Denken erkannt, ergriffen. Wäre das Erkanntwerden ein Entstehen, so könnten wir nichts Positives von ihm aussagen in Bezug auf eine Zeit, die diesem vorgeblichen Entstehen vorherginge.

Der Raum gehört nach Kant der Erscheinung an. Es wäre möglich, dass er andern Vernunftwesen sich ganz anders als uns darstellte. Ja, wir können nicht einmal wissen, ob er dem einen Menschen so wie dem andern erscheint; denn wir können die Raumanschauung des einen nicht neben die des andern legen, um sie zu vergleichen. Aber dennoch ist darin etwas Objectives enthalten; Alle erkennen dieselben geometrischen Axiome, wenn auch nur durch die That, an und müssen es, um sich in der Welt zurechtzufinden. Objectiv ist darin das Gesetzmässige, Begriffliche, Beurtheilbare, was sich in Worten ausdrücken lässt. Das rein Anschauliche ist nicht mittheilbar. Nehmen wir zur Verdeutlichung zwei Vernunftwesen an, denen nur die projectivischen Eigenschaften und Beziehungen anschaulich sind: das Liegen von drei Punkten in einer Gerade, von vier Punkten in einer Ebene u. s. w.; es möge dem einen das als Ebene erscheinen, was das andere als Punkt anschaut und umgekehrt. Was dem einen die Verbindungslinie von Punkten ist, möge dem andern die Schnittkante von Ebenen sein u. s. w. immer dualistisch entsprechend. Dann könnten sie sich sehr wohl mit einander verständigen und würden die Verschiedenheit ihres Anschauens nie gewahr werden, weil in der projectivischen Geometrie jedem Lehrsatze ein anderer dualistisch gegenübersteht; denn das Abweichen in einer ästhetischen

I distinguish what I call objective from what is handleable or spatial or actual. The axis of the earth is objective, so is the centre of mass of the solar system, but I should not call them actual in the way the earth itself is so. We often speak of the equator as an *imaginary* line; but it would be wrong to call it a *fictitious* line; it is not a creature of thought, the product of a psychological process, but is only recognized or apprehended by thought. If to be recognized were to be created, then we should be able to say nothing positive about the equator for any period earlier than the date of its alleged creation.

Space, according to KANT, belongs to appearance. For other rational beings it might take some form quite different from that in which we know it. Indeed, we cannot even know whether it appears the same to one man as to another; for we cannot, in order to compare them, lay one man's intuition of space beside another's. Yet there is something objective in it all the same; everyone recognizes the same geometrical axioms, even if only by his behaviour, and must do so if he is to find his way about the world. What is objective in it is what is subject to laws, what can be conceived and judged, what is expressible in words. What is purely intuitable is not communicable. To make this clear, let us suppose two rational beings such that projective properties and relations are all they can intuite—the lying of three points on a line, of four points on a plane, and so on; and let what the one intuites as a plane appear to the other as a point, and vice versa, so that what for the one is the line joining two points for the other is the line of intersection of two planes, and so on with the one intuition always dual to the other. In these circumstances they could understand one another quite well and would never realize the difference between their intuitions, since in projective geometry every proposition has its dual counterpart; any disagreements over points of aesthetic appreciation would not

Werthschätzung würde kein sicheres Zeichen sein. In Bezug auf alle geometrische Lehrsätze wären sie völlig im Einklange; sie würden sich nur die Wörter in ihre Anschauung verschieden übersetzen. Mit dem Worte „Punkt" verbände etwa das eine diese, das andere jene Anschauung. So kann man immerhin sagen, dass ihnen dies Wort etwas Objectives bedeute; nur darf man unter dieser Bedeutung nicht das Besondere ihrer Anschauung verstehn. Und in diesem Sinne ist auch die Erdaxe objectiv.

Man denkt gewöhnlich bei „weiss" an eine gewisse Empfindung, die natürlich ganz subjectiv ist; aber schon im gewöhnlichen Sprachgebrauche, scheint mir, tritt ein objectiver Sinn vielfach hervor. Wenn man den Schnee weiss nennt, so will man eine objective Beschaffenheit ausdrücken, die man beim gewöhnlichen Tageslicht an einer gewissen Empfindung erkennt. Wird er farbig beleuchtet, so bringt man das bei der Beurtheilung in Anschlag. Man sagt vielleicht: er erscheint jetzt roth, aber er ist weiss. Auch der Farbenblinde kann von roth und grün reden, obwohl er diese Farben in der Empfindung nicht unterscheidet. Er erkennt den Unterschied daran, dass Andere ihn machen, oder vielleicht durch einen physikalischen Versuch. So bezeichnet das Farbenwort oft nicht unsere subjective Empfindung, von der wir nicht wissen können, dass sie mit der eines Andern übereinstimmt — denn offenbar verbürgt das die gleiche Benennung keineswegs — sondern eine objective Beschaffenheit. So verstehe ich unter Objectivität eine Unabhängigkeit von unserm Empfinden, Anschauen und Vorstellen, von dem Entwerfen innerer Bilder aus den Erinnerungen früherer Empfindungen, aber nicht eine Unabhängigkeit von der Vernunft; denn die Frage beantworten, was die Dinge unabhängig von der Vernunft sind, hiesse urtheilen, ohne zu urtheilen, den Pelz waschen, ohne ihn nass zu machen.

§ 27. Deswegen kann ich auch Schloemilch*) nicht

---

*) Handbuch der algebraischen Analysis, S. 1.

be conclusive evidence. Over all geometrical theorems they would be in complete agreement, only interpreting the words differently in terms of their respective intuitions. With the word "point", for example, one would connect one intuition and the other another. We can therefore still say that this word has for them an objective meaning, provided only that by this meaning we do not understand any of the peculiarities of their respective intuitions. And in this sense the axis of the earth too is objective.

The word "white" ordinarily makes us think of a certain sensation, which is, of course, entirely subjective; but even in ordinary everyday speech, it often bears, I think, an objective sense. When we call snow white, we mean to refer to an objective quality which we recognize, in ordinary daylight, by a certain sensation. If the snow is being seen in a coloured light, we take that into account in our judgement and say, for instance, "It *appears* red at present, but it *is* white." Even a colour-blind man can speak of red and green, in spite of the fact that he does not distinguish between these colours in his sensations; he recognizes the distinction by the fact that others make it, or perhaps by making a physical experiment. Often, therefore, a colour word does not signify our subjective sensation, which we cannot know to agree with anyone else's (for obviously our calling things by the same name does not guarantee as much), but rather an objective quality. It is in this way that I understand objective to mean what is independent of our sensation, intuition and imagination, and of all construction of mental pictures out of memories of earlier sensations, but not what is independent of the reason,—for what are things independent of the reason? To answer that would be as much as to judge without judging, or to wash the fur without wetting it.

§ 27. For that reason I cannot agree with SCHLOEMILCH[1]

---

[1] *Handbuch der algebraischen Analysis*, p. 1.

zustimmen, der die Zahl Vorstellung der Stelle eines Objects in einer Reihe nennt*). Wäre die Zahl eine Vorstellung, so wäre die Arithmetik Psychologie. Das ist sie so wenig, wie etwa die Astronomie es ist. Wie sich diese nicht mit den Vorstellungen der Planeten, sondern mit den Planeten selbst beschäftigt, so ist auch der Gegenstand der Arithmetik keine Vorstellung. Wäre die Zwei eine Vorstellung, so wäre es zunächst nur die meine. Die Vorstellung eines Andern ist schon als solche eine andere. Wir hätten dann vielleicht viele Millionen Zweien. Man müsste sagen: meine Zwei, deine Zwei, eine Zwei, alle Zweien. Wenn man latente oder unbewusste Vorstellungen annimmt, so hätte man auch unbewusste Zweien, die dann später wieder bewusste würden. Mit den heranwachsenden Menschen entständen immer neue Zweien, und wer weiss, ob sie sich nicht in Jahrtausenden so veränderten, dass $2 \times 2 = 5$ würde. Trotzdem wäre

*) Man kann dagegen auch einwenden, dass dann immer dieselbe Vorstellung einer Stelle erscheinen müsste, wenn dieselbe Zahl auftritt, was offenbar falsch ist. Das Folgende würde nicht zutreffen, wenn er unter Vorstellung eine objective Idee verstehen wollte; aber welcher Unterschied wäre dann zwischen der Vorstellung der Stelle und der Stelle selbst?

Die Vorstellung im subjectiven Sinne ist das, worauf sich die psychologischen Associationsgesetze beziehen; sie ist von sinnlicher, bildhafter Beschaffenheit. Die Vorstellung im objectiven Sinne gehört der Logik an und ist wesentlich unsinnlich, obwohl das Wort, welches eine objective Vorstellung bedeutet, oft auch eine subjective mit sich führt, die jedoch nicht seine Bedeutung ist. Die subjective Vorstellung ist oft nachweisbar verschieden in verschiedenen Menschen, die objective für alle dieselbe. Die objectiven Vorstellungen kann man eintheilen in Gegenstände und Begriffe. Ich werde, um Verwirrung zu vermeiden, „Vorstellung" nur im subjectiven Sinne gebrauchen. Dadurch, dass Kant mit diesem Worte beide Bedeutungen verband, hat er seiner Lehre eine sehr subjective, idealistische Färbung gegeben und das Treffen seiner wahren Meinung erschwert. Die hier gemachte Unterscheidung ist so berechtigt wie die zwischen Psychologie und Logik. Möchte man diese immer recht streng auseinanderhalten!

either, when he calls number the idea of the position of an item in a series.[1] If number were an idea, then arithmetic would be psychology. But arithmetic is no more psychology than, say, astronomy is. Astronomy is concerned, not with ideas of the planets, but with the planets themselves, and by the same token the objects of arithmetic are not ideas either. If the number two were an idea, then it would have straight away to be private to me only. Another man's idea is, *ex vi termini*, another idea. We should then have it might be many millions of twos on our hands. We should have to speak of my two and your two, of one two and all twos. If we accept latent or unconscious ideas, we should have unconscious twos among them, which would then return subsequently to consciousness. As new generations of children grew up, new generations of twos would continually be being born, and in the course of millennia these might evolve, for all we could tell, to such a pitch that two of them would make five. Yet, in spite of all this, it

[1] Another possible objection is, that on this theory the same idea of a position in a series would have always to appear whenever the same number occurred, which obviously does not happen. My arguments would be beside the point if he meant by idea an objective notion [*Idee*]; but in that case what distinction would there be between the idea of the position and the position itself?

An idea in the subjective sense is what is governed by the psychological laws of association; it is of a sensible, pictorial character. An idea in the objective sense belongs to logic and is in principle non-sensible, although the word which means an objective idea is often accompanied by a subjective idea, which nevertheless is not its meaning. Subjective ideas are often demonstrably different in different men, objective ideas are the same for all. Objective ideas can be divided into objects and concepts. I shall myself, to avoid confusion, use "idea" only in the subjective sense. It is because Kant associated both meanings with the word that his doctrine assumed such a very subjective, idealist complexion, and his true view was made so difficult to discover. The distinction here drawn stands or falls with that between psychology and logic. If only these themselves were to be kept always rigidly distinct!

es zweifelhaft, ob es, wie man gewöhnlich meint, unendlich viele Zahlen gäbe. Vielleicht wäre 10^{10} nur ein leeres Zeichen, und es gäbe gar keine Vorstellung, in irgendeinem Wesen, die so benannt werden könnte.

Wir sehen, zu welchen Wunderlichkeiten es führt, wenn man den Gedanken etwas weiter ausspinnt, dass die Zahl eine Vorstellung sei. Und wir kommen zu dem Schlusse, dass die Zahl weder räumlich und physikalisch ist, wie Mills Haufen von Kieselsteinen und Pfeffernüssen, noch auch subjectiv wie die Vorstellungen, sondern unsinnlich und objectiv. Der Grund der Objectivität kann ja nicht in dem Sinneseindrucke liegen, der als Affection unserer Seele ganz subjectiv ist, sondern soweit ich sehe, nur in der Vernunft.

Es wäre wunderbar, wenn die allerexacteste Wissenschaft sich auf die noch zu unsicher tastende Psychologie stützen sollte.

Die Anzahl als Menge.

§ 28. Einige Schriftsteller erklären die Anzahl als eine Menge, Vielheit oder Mehrheit. Ein Uebelstand besteht hierbei darin, dass die Zahlen o und 1 von dem Begriffe ausgeschlossen werden. Jene Ausdrücke sind recht unbestimmt: bald nähern sie sich mehr der Bedeutung von „Haufe," „Gruppe," „Aggregat" — wobei an ein räumliches Zusammensein gedacht wird — bald werden sie fast gleichbedeutend mit „Anzahl" gebraucht, nur unbestimmter. Eine Auseinanderlegung des Begriffes der Anzahl kann darum in einer solchen Erklärung nicht gefunden werden. Thomae*) verlangt zur Bildung der Zahl, dass verschiedenen Objectenmengen verschiedene Namen gegeben werden. Damit ist offenbar eine schärfere Bestimmung jener Objectenmengen gemeint, für welche die Namengebung nur das äussere Zeichen ist. Welcher Art nun diese Bestimmung sei, das ist die

---

*) Elementare Theorie der analytischen Functionen, S. 1.

would still be doubtful whether there existed infinitely many numbers, as we ordinarily suppose. 10^{10}, perhaps, might be only an empty symbol, and there might exist no idea at all, in any being whatever, to answer to the name.

Weird and wonderful, as we see, are the results of taking seriously the suggestion that number is an idea. And we are driven to the conclusion that number is neither spatial and physical, like MILL's piles of pebbles and gingersnaps, nor yet subjective like ideas, but non-sensible and objective. Now objectivity cannot, of course, be based on any sense-impression, which as an affection of our mind is entirely subjective, but only, so far as I can see, on the reason.

It would be strange if the most exact of all the sciences had to seek support from psychology, which is still feeling its way none too surely.

Numbers as sets.

§ 28. Some writers define Number as a set or multitude or plurality. All these views suffer from the drawback that the concept will not then cover the numbers 0 and 1. Moreover, these terms are utterly vague: sometimes they approximate in meaning to "heap" or "group" or "agglomeration", referring to a juxtaposition in space, sometimes they are so used as to be practically equivalent to "Number", only vaguer. No analysis of the concept of Number, therefore, is to be found in a definition of this kind. THOMAE[1] requires for the formation of number that item-sets which differ be given different names. By this he evidently means to refer to a process of bringing out more sharply the characteristics of the sets in question, of which the giving of names is only the external sign. The question is, just what is this process like?

---

[1] *Elementare Theorie der analytischen Functionen*, p. 1.

Frage. Es würde offenbar die Idee der Zahl nicht entstehen, wenn man· für „3 Sterne," „3 Finger," „7 Sterne" Namen einführen wollte, in denen keine gemeinsamen Bestandtheile erkennbar wären. Es kommt nicht darauf an, dass überhaupt Namen gegeben werden, sondern dass für sich bezeichnet werde, was Zahl daran ist. Dazu ist nöthig, dass es in seiner Besonderheit erkannt sei.

Noch ist folgende Verschiedenheit zu beachten. Einige nennen die Zahl eine Menge von Dingen oder Gegenständen; Andere wie schon Euklid*), erklären sie als eine Menge von Einheiten. Dieser Ausdruck bedarf einer besondern Erörterung.

III. Meinungen über Einheit und Eins.

Drückt das Zahlwort „Ein" eine Eigenschaft von Gegenständen aus?

§ 29. In den Definitionen, die Euklid am Anfange des 7. Buches der Elemente giebt, scheint er mit dem Worte „μονάς" bald einen zu zählenden Gegenstand, bald eine Eigenschaft eines solchen, bald die Zahl Eins zu bezeichnen. Ueberall kommt man mit der Uebersetzung „Einheit" durch, aber nur, weil dies Wort selbst in diesen verschiedenen Bedeutungen schillert.

Schröder**) sagt: „Jedes der zu zählenden Dinge wird Einheit genannt." Es fragt sich, weshalb man die Dinge erst unter den Begriff der Einheit bringt und nicht einfach erklärt: Zahl ist eine Menge von Dingen, womit wir wieder auf das Vorige zurückgeworfen wären. Man könnte zunächst in der Benennung der Dinge als Einheiten eine nähere Bestimmung finden wollen, indem man der sprachlichen Form folgend „Ein" als Eigenschaftswort ansieht

---

*) 7. Buch der Elemente im Anfange: Μονάς ἐστι, καθ' ἣν ἕκαστον τῶν ὄντων ἓν λέγεται. 'Αριθμὸς δὲ τὸ ἐκ μονάδων συγκείμενον πλῆθος.

**) A. a. O. S. 5.

Obviously, the notion of number would not result if, instead of "3 stars", "3 fingers" and "7 stars", we tried introducing names in which there were no recognizable common elements. It is not a matter simply of assigning names, but of symbolizing in its own right the numerical element. For this, we must needs have come to recognize that element in its peculiarity.

Furthermore, it should be noted that there are two different views. Some call number a set of things or objects; others, following EUCLID[1], define it as a set of units. This last expression demands a separate discussion.

III. Views on unity and one*.

Does the number word "one" stand for a property of objects?

§ 29. In the definitions which EUCLID gives at the beginning of Book VII of the Elements, he seems to mean by the word "μονάς" sometimes an object to be counted, sometimes a property of such an object, and sometimes the number one. We can translate it consistently by the German "Einheit", but only because that word itself shifts over the same variety of meanings.

According to SCHRÖDER[2]: "Each of the things to be counted is called a unit." We may well wonder why we must first bring the things under the concept of unity, instead of simply defining number right away as a set of things, which would throw us back once again onto the first of the two views. The most obvious answer is that in calling the things units we are supposed to be adding to our description of them; under the influence of the grammatical form, we are regarding "one"

---

[1] Μονάς ἐστι, καθ' ἣν ἕκαστον τῶν ὄντων ἓν λέγεται. Ἀριθμὸς δὲ τὸ ἐκ μονάδων συγκείμενον πλῆθος. ["A unit is that by virtue of which each of the things that exist is called one. A number is a multitude composed of units."]

[2] Op. cit., p. 5.

* [It is not possible in English to do entire justice to the ambiguities of the German *Einheit*, which covers both "unit" and "unity", not to mention "oneness". Moreover *Einheit* is a verbal derivative of *Ein* (whereas "unit/y" is not derived directly from "one"), and derivatives of either word can be described alike as derivatives of *Ein* or "one". These facts make §§ 29, 32 and 37, in particular, more plausible in German than in English.]

und „Eine Stadt" so auffasst wie „weiser Mann". Dann würde eine Einheit ein Gegenstand sein, dem die Eigenschaft „Ein" zukäme und würde sich zu „Ein" ähnlich verhalten wie „ein Weiser" zu dem Adjectiv „weise". Zu den Gründen, die oben dagegen geltend gemacht sind, dass die Zahl eine Eigenschaft von Dingen sei, treten hier noch einige besondere hinzu. Auffallend wäre zunächst, dass jedes Ding diese Eigenschaft hätte. Es wäre unverständlich, weshalb man überhaupt noch einem Dinge ausdrücklich die Eigenschaft beilegt. Nur durch die Möglichkeit, dass etwas nicht weise sei, gewinnt die Behauptung, Solon sei weise, einen Sinn. Der Inhalt eines Begriffes nimmt ab, wenn sein Umfang zunimmt; wird dieser allumfassend, so muss der Inhalt ganz verloren gehen. Es ist nicht leicht zu denken, wie die Sprache dazu käme, ein Eigenschaftswort zu schaffen, das gar nicht dazu dienen könnte, einen Gegenstand näher zu bestimmen.

Wenn „Ein Mensch" ähnlich wie „weiser Mensch" aufzufassen wäre, so sollte man denken, dass „Ein" auch als Praedicat gebraucht werden könnte, sodass man wie „Solon war weise" auch sagen könnte „Solon war Ein" oder „Solon war Einer". Wenn nun der letzte Ausdruck auch vorkommen kann, so ist er doch für sich allein nicht verständlich. Er kann z. B. heissen: Solon war ein Weiser, wenn „Weiser" aus dem Zusammenhange zu ergänzen ist. Aber allein scheint „Ein" nicht Praedicat sein zu können*). Noch deutlicher zeigt sich dies beim Plural. Während man „Solon war weise" und „Thales war weise" zusammenziehen kann in „Solon und Thales waren weise," kann man nicht sagen „Solon und Thales waren Ein". Hiervon wäre die

---

*) Es kommen Wendungen vor, die dem zu widersprechen scheinen; aber bei genauerer Betrachtung wird man finden, dass ein Begriffswort zu ergänzen ist, oder dass „Ein" nicht als Zahlwort gebraucht wird, dass nicht die Einzigkeit, sondern die Einheitlichkeit behauptet werden soll.

as a word for a property and taking "one city" in the same way as "wise man". In that case a unit would be an object characterized by the property "one" and would stand to "one" in the same relation as "a sage" to the adjective "wise". Now reasons have already been given as conclusive against the view that number is a property of things; but there are several further arguments against the present suggestion in particular. It must strike us immediately as remarkable that every single thing should possess this property. It would be incomprehensible why we should still ascribe it expressly to a thing at all. It is only in virtue of the possibility of something not being wise that it makes sense to say "Solon is wise." The content of a concept diminishes as its extension increases; if its extension becomes all-embracing, its content must vanish altogether. It is not easy to imagine how language could have come to invent a word for a property which could not be of the slightest use for adding to the description of any object whatsoever.

If it were correct to take "one man" in the same way as "wise man", we should expect to be able to use "one" also as a grammatical predicate, and to be able to say "Solon was one" just as much as "Solon was wise". It is true that "Solon was one" can actually occur, but not in a way to make it intelligible on its own in isolation. It may, for example, mean "Solon was a wise man", if "wise man" can be supplied from the context. In isolation, however, it seems that "one" cannot be a predicate.[1] This is even clearer if we take the plural. Whereas we can combine "Solon was wise" and "Thales was wise" into "Solon and Thales were wise", we cannot say "Solon and Thales were one". But it is hard to see why this

---

[1] Usages do occur which appear to contradict this; but if we look more closely we shall find that some general term has to be supplied, or else that "one" is not being used as a number word—that what it is intended to assert is the character (not of being unique, but) of being unitary.

41

Unmöglichkeit nicht einzusehen, wenn „Ein" sowie „weise" eine Eigenschaft sowohl des Solon als auch des Thales wäre.

§ 30. Damit hangt es zusammen, dass man keine Definition der Eigenschaft „Ein" hat geben können. Wenn Leibniz*) sagt: „Eines ist, was wir durch Eine That des Verstandes zusammenfassen", so erklärt er „Ein" durch sich selbst. Und können wir nicht auch Vieles durch Eine That des Verstandes zusammenfassen? Dies wird von Leibniz an derselben Stelle zugestanden. Aehnlich sagt Baumann**): „Eines ist, was wir als Eines auffassen" und weiter: „Was wir als Punkt setzen oder nicht mehr als getheilt setzen wollen, das sehen wir als Eines an; aber jedes Eins der äussern Anschauung, der reinen wie der empirischen, können wir auch als Vieles ansehen. Jede Vorstellung ist Eine, wenn abgegränzt gegen eine andere Vorstellung; aber in sich kann sie wieder in Vieles unterschieden werden." So verwischt sich jede sachliche Begrenzung des Begriffes und alles hangt von unserer Auffassung ab. Wir fragen wieder: welchen Sinn kann es haben, irgendeinem Gegenstande die Eigenschaft „Ein" beizulegen, wenn je nach der Auffassung jeder Einer sein und auch nicht sein kann? Wie kann auf einem so verschwommenen Begriffe eine Wissenschaft beruhen, die grade in der grössten Bestimmtheit und Genauigkeit ihren Ruhm sucht?

§ 31. Obwohl nun Baumann***) den Begriff der Eins auf innerer Anschauung beruhen lässt, so nennt er doch in der eben angeführten Stelle als Merkmale die Ungetheiltheit und die Abgegränztheit. Wenn diese zuträfen, so wäre zu erwarten, dass auch Thiere eine gewisse Vorstellung von Einheit haben könnten. Ob wohl ein Hund beim Anblick des Mondes eine wenn auch noch so unbestimmte Vorstellung

---

*) Baumann a. a. O. Bd. II. S. 2; Erdm. S. 8.
**) A. a. O. Bd. II. S. 669.
***) A. a. O. Bd. II. S. 669.

should be impossible, if "one" were a property both of Solon and of Thales in the same way that "wise" is.

§ 30. In line with this is the fact that no one has ever been able to give a definition of the property "one". LEIBNIZ[1] indeed says that "By *one* is meant whatever we grasp in one act of the understanding," but this is to define "one" in terms of itself. Besides, surely we can also grasp what is many in one act of the understanding? LEIBNIZ admits as much in the same passage. BAUMANN[2] does no better when he says: "That is one, which we apprehend as one", and further: "Whatever we take as a point, or refuse to take as further subdivided into parts, that we regard as one; but every one of outer intuition, whether empirical or pure, can also be regarded as a many. Every idea is one when isolated in contrast with another; but in itself it can again be distinguished into a many." This sweeps away every limit to the application of the concept imposed by the nature of the facts, and everything is made dependent on our way of regarding them. I ask once more: How can it make sense to ascribe the property "one" to any object whatever, when every object, according as to how we look at it, can be either one or not one? How can a science which bases its claim to fame precisely on being as definite and accurate as possible repose on a concept as hazy as this is?

§ 31. Now although BAUMANN[3] bases the concept of one on inner intuition, he refers nevertheless, in the passage just cited, to certain criteria for being one, namely being undivided and being isolated. If this were correct, then we should have to expect animals, too, to be capable of having some sort of idea of unity. Can it be that a dog staring at the moon does have an idea, however ill-defined, of what we signify by the

---

[1] Baumann, op. cit., Vol. II, p. 2 (Erdmann edn., p. 8).

[2] Op. cit., Vol. II, p. 669.

[3] Op. cit., Vol. II, p. 669.

von dem hat, was wir mit dem Worte „Ein" bezeichen? Schwerlich! Und doch unterscheidet er gewiss einzelne Gegenstände: ein andrer Hund, sein Herr, ein Stein, mit dem er spielt, erscheinen ihm gewiss ebenso abgegrenzt, für sich bestehend, ungetheilt wie uns. Zwar wird er einen Unterschied merken, ob er sich gegen viele Hunde zu vertheidigen hat oder nur gegen Einen, aber dies ist der von Mill physikalisch genannte Unterschied. Es käme darauf besonders an, ob er von dem Gemeinsamen, welches wir durch das Wort „Ein" ausdrücken, ein wenn auch noch so dunkles Bewusstsein hat z. B. in den Fällen, wo er von Einem grössern Hunde gebissen wird, und wo er Eine Katze verfolgt. Das ist mir unwahrscheinlich. Ich folgere daraus, dass die Idee der Einheit nicht, wie Locke*) meint, dem Verstande durch jenes Object draussen und jede Idee innen zugeführt, sondern von uns durch die höhern Geisteskräfte erkannt wird, die uns vom Thiere unterscheiden. Dann können solche Eigenschaften der Dinge wie Ungetheiltheit und Abgegrenztheit, die von den Thieren ebenso gut wie von uns bemerkt werden, nicht das Wesentliche an unserm Begriffe sein.

§ 32. Doch kann man einen gewissen Zusammenhang vermuthen. Darauf deutet die Sprache hin, indem sie von „Ein" „einig" ableitet. Etwas ist desto mehr geeignet, als besonderer Gegenstand aufgefasst zu werden, je mehr die Unterschiede in ihm gegenüber den Unterschieden von der Umgebung zurücktreten, je mehr der innere Zusammenhang den mit der Umgebung überwiegt. So bedeutet „einig" eine Eigenschaft, die dazu veranlasst, etwas in der Auffassung von der Umgebung abzusondern und für sich zu betrachten. Wenn das französische „uni" „eben," „glatt" heisst, so ist dies so zu erklären. Auch das Wort „Einheit" wird in ähnlicher Weise gebraucht, wenn von politischer Einheit

*) Baumann a. a. O. Bd. I. S. 409.

word "one"? This is hardly credible—and yet it certainly
distinguishes individual objects: another dog, its master, a
stone it is playing with, these certainly appear to the dog every
bit as isolated, as self-contained, as undivided, as they do to us.
It will notice a difference, no doubt, between being set on by
several other dogs and being set on by only one, but this is
what MILL calls the physical difference. We need to know
specifically: is the dog conscious, however dimly, of that
common element in the two situations which we express by the
word "one", when, for example, it first is bitten by one larger
dog and then chases one cat? This seems to me unlikely. I
infer, therefore, that the notion of unity is not, as LOCKE[1]
holds, "suggested to the understanding by every object without us, and every idea within", but becomes known to us
through the exercise of those higher intellectual powers which
distinguish men from brutes. Consequently, such properties
of things as being undivided or being isolated, which animals
perceive quite as well as we do, cannot be what is essential in
our concept.

§ 32. Still, we may suspect that they have some sort of
connexion with it. Language indicates as much by forming
"united" as a derivative from "one". The more the internal
contrasts within a thing fade into insignificance by comparison
with the contrasts between it and its environment, and the
more the internal connexions among its elements overshadow
its connexions with its environment, the more natural it
becomes for us to regard it as a distinct object. For a thing to
be "united" means that it has a property which causes us, when
we think of it, to sever it from its environment and consider it
on its own. In the same way we can explain how the French
"uni" comes to mean "even" or "smooth". The word "unity"
too, is used in a similar manner, when we speak of the political

---

[1] Baumann, op. cit., Vol. I, p. 409 [*Essay*, Bk. II, cap. vii, § 7].

eines Landes, Einheit eines Kunstwerks gesprochen wird*).
Aber in diesem Sinne gehört „Einheit" weniger zu „Ein"
als zu „einig" oder „einheitlich." Denn, wenn man sagt,
die Erde habe Einen Mond, so will man diesen damit nicht
für einen abgegrenzten, für sich bestehenden, ungetheilten
Mond erklären; sondern man sagt dies im Gegensatze zu
dem, was bei der Venus, dem Mars oder dem Jupiter vorkommt. In Bezug auf Abgegrenztheit und Ungetheiltheit
könnten sich die Monde des Jupiter wohl mit unserm messen
und sind in dem Sinne ebenso einheitlich.

§ 33. Die Ungetheiltheit wird von einigen Schriftstellern bis zur Untheilbarkeit gesteigert. G. Köpp**) nennt
jedes unzerlegbar und für sich bestehend gedachte sinnlich
oder nicht sinnlich wahrnehmbare Ding ein Einzelnes und
die zu zählenden Einzelnen Einse, wo offenbar „Eins" in
dem Sinne von „Einheit" gebraucht wird. Indem Baumann
seine Meinung, die äussern Dinge stellten keine strengen
Einheiten dar, damit begründet, dass wir die Freiheit hätten,
sie als Vieles zu betrachten, giebt auch er die Unzerlegbarkeit für ein Merkmal der strengen Einheit aus. Dadurch
dass man den innern Zusammenhang bis zum Unbedingten
steigert, will man offenbar ein Merkmal der Einheit gewinnen,
das von der willkührlichen Auffassung unabhängig ist. Dieser
Versuch scheitert daran, dass dann fast nichts übrig bliebe,
was Einheit genannt und gezählt werden dürfte. Deshalb
wird auch sofort der Rückzug damit angetreten, dass man
nicht die Unzerlegbarkeit selbst, sondern das als unzerlegbar
Gedachtwerden als Merkmal aufstellt. Damit ist man denn
bei der schwankenden Auffassung wieder angekommen. Und
wird denn dadurch etwas gewonnen, dass man sich die Sachen
anders denkt als sie sind? Im Gegentheil! aus einer falschen

---

*) Ueber die Geschichte des Wortes „Einheit" vergl. Eucken, Ge-
schichte der philosophischen Terminologie. S. 122—3, S. 136, S. 220.
**) Schularithmetik. Eisenach 1867. S. 5 u. 6.

unity of a country or the unity of a work of art.[1] But in this sense "unity" is connected not so much with "one" as with "united" or "unitary". For when we say the Earth has one moon, we do not mean to point out that our satellite is isolated, or self-contained, or undivided; rather, we are contrasting the satellite system of the Earth with that of Venus or Mars or Jupiter. So far as being isolated goes, or being undivided, the moons of Jupiter could stand up quite well to our moon, and in that sense they are every bit as unitary.

§ 33. Some writers go still further, and demand something not merely undivided but indivisible. G. Köpp[2] calls whatever is thought of as self-contained and incapable of dissection, whether we become acquainted with it through the senses or otherwise, an individual; and individuals which are to be numbered he calls ones, evidently using "one" here in the sense of "unit". BAUMANN too, when arguing for his view that external things do not present us with strict units on the ground that we are free to treat them as many, gives as a criterion of a strict unit that it must be incapable of dissection. Obviously, by tightening up its internal cohesion without limit, they hope to arrive at a criterion for their unit which is independent of any arbitrary way of regarding things. This attempt collapses because we are then left with practically nothing fit to be called a unit and to be numbered. The result is that we at once begin to retrace our steps, by giving as the criterion not that the thing itself should be incapable of dissection in fact, but that we should think of it as such. This brings us back once again to our way of regarding things, with all its fluctuations. And is it really of any advantage to think of things as being what they are not? On the contrary, any

---

[1] On the history of the word "unit" cf. Eucken, *Geschichte der philosophischen Terminologie*, pp. 122–23, 136, 220.

[2] *Schularithmetik*, pp. 5–6, Eisenach 1867.

Annahme können falsche Folgerungen fliessen. Wenn man aber aus der Unzerlegbarkeit nichts schliessen will, was nützt sie dann? wenn man von der Strenge des Begriffes ohne Schaden etwas ablassen kann, ja es sogar muss, wozu dann diese Strenge? Aber vielleicht soll man an die Zerlegbarkeit nur nicht denken. Als ob durch Mangel an Denken etwas erreicht werden könnte! Es giebt aber Fälle, wo man gar nicht vermeiden kann, an die Zerlegbarkeit zu denken, wo sogar ein Schluss auf der Zusammensetzung der Einheit beruht, z. B. bei der Aufgabe: Ein Tag hat 24 Stunden, wieviel Stunden haben 3 Tage?

Sind die Einheiten einander gleich?

§ 34. So misslingt denn jeder Versuch, die Eigenschaft „Ein" zu erklären, und wir müssen wohl darauf verzichten, in der Bezeichnung der Dinge als Einheiten eine nähere Bestimmung zu sehen. Wir kommen wieder auf unsere Frage zurück: weshalb nennt man die Dinge Einheiten, wenn „Einheit" nur ein andrer Name für Ding ist, wenn alle Dinge Einheiten sind oder als solche aufgefasst werden können? E. Schröder*) giebt als Grund die den Objecten der Zählung zugeschriebene Gleichheit an. Zunächst ist nicht zu sehen, warum die Wörter „Ding" und „Gegenstand" dies nicht ebenso gut andeuten könnten. Dann fragt es sich: weshalb wird den Gegenständen der Zählung Gleichheit zugeschrieben? Wird sie ihnen nur zugeschrieben, oder sind sie wirklich gleich? Jedenfalls sind nie zwei Gegenstände durchaus gleich. Andrerseits kann man wohl fast immer eine Hinsicht ausfindig machen, in der zwei Gegenstände übereinstimmen. So sind wir wieder bei der willkührlichen Auffassung angelangt, wenn wir nicht gegen die Wahrheit den Dingen eine weitergehende Gleichheit zuschreiben wollen, als ihnen zukommt. In der That nennen viele Schriftsteller

*) A. a. O. S. 5.

conclusions drawn from a false assumption are liable to be false; while if there are no conclusions to be drawn from our unit's being incapable of dissection, why bother to assume that it is? If it does no harm, and in fact is actually necessary, to take our strict units none too strictly, what was the point of being strict? Unless perhaps all that was meant was that we should not think of the possibilities of dissection—as though lack of thought could get us anywhere! Besides, there are cases where we simply cannot avoid thinking of them, where a conclusion is actually based on the way in which a unit is made up of parts, as for instance in the problem: If there are 24 hours in one day, how many are there in three days?

Are units identical with one another?*

§ 34. Every attempt to define "one" as a property having thus failed, we must finally abandon the view that in designating a thing a unit we are adding to our description of it. We come back once again to our question: Why do we call things units, if "unit" is only another name for thing, if any and every thing is a unit or can be regarded as one? E. SCHRÖDER[1] gives as the reason, that the word is used for ascribing to the items that are to be numbered the necessary identity. But to begin with it is not easy to see why the words "thing" and "object" could not indicate this just as well. And further, it is natural to ask: Why do we ascribe identity to objects that are to be numbered? And is it only ascribed to them, or are they really identical? In any case, no two objects are *ever* completely identical. On the other hand, of course, we can practically always engineer some respect in which any two objects whatever agree. And with that we are back once more at our arbitrary way of regarding things, unless we are willing, regardless of truth, to ascribe to things an identity going beyond any they actually possess. In actual fact, many writers

[1] Op. cit., p. 5.

* [*gleich*. See footnote on p. IIe above.]

die Einheiten ohne Einschränkung gleich. Hobbes*) sagt: „Die Zahl, absolut gesagt, setzt in der Mathematik unter sich gleiche Einheiten voraus, aus denen sie hergestellt wird." Hume**) hält die zusammensetzenden Theile der Quantität und Zahl für ganz gleichartig. Thomae***) nennt ein Individuum der Menge Einheit und sagt: „Die Einheiten sind einander gleich." Ebenso gut oder vielmehr richtiger könnte man sagen: die Individuen der Menge sind von einander verschieden. Was hat nun diese vorgebliche Gleichheit für die Zahl zu bedeuten? Die Eigenschaften, durch die sich die Dinge unterscheiden, sind für ihre Anzahl etwas Gleichgiltiges und Fremdes. Darum will man sie fern halten. Aber das gelingt in dieser Weise nicht. Wenn man, wie Thomae verlangt, „von den Eigenthümlichkeiten der Individuen einer Objectenmenge abstrahirt" oder „bei der Betrachtung getrennter Dinge von den Merkmalen absieht, durch welche sich die Dinge unterscheiden," so bleibt nicht, wie Lipschitz meint, „der Begriff der Anzahl der betrachteten Dinge" zurück, sondern man erhält einen allgemeinen Begriff, unter den jene Dinge fallen. Diese selbst verlieren dadurch nichts von ihren Besonderheiten. Wenn ich z. B. bei der Betrachtung einer weissen und einer schwarzen Katze von den Eigenschaften absehe, durch die sie sich unterscheiden, so erhalte ich etwa den Begriff „Katze". Wenn ich nun auch beide unter diesen Begriff bringe und sie etwa Einheiten nenne, so bleibt die weisse doch immer weiss und die schwarze schwarz. Auch dadurch, dass ich an die Farben nicht denke oder mir vornehme, keine Schlüsse aus deren Verschiedenheit zu ziehen, werden die Katzen nicht farblos und bleiben ebenso verschieden, wie sie waren. Der Begriff „Katze," der durch

*) Baumann a. a. O. Bd. I. S. 242.
**) Ebenda Bd. II. S. 568.
***) A. a. O. S. 1.

do call units identical without any qualification. HOBBES[1] states that: "Number in the absolute sense in mathematics presupposes units identical* one with another out of which it is formed." HUME[2] holds the component parts of quantity and number to be entirely similar. THOMAE[3] calls the individual member of his set a unit, and says in so many words that units are identical with each other—though we might with as much or even more justice say that the individual members of his set must be different from each other. Now what has this alleged identity to do with number? The properties which serve to distinguish things from one another are, when we are considering their Number, immaterial and beside the point. That is why we want to keep them out of it. But we shall not succeed along the present lines. For suppose that we do, as THOMAE demands, "abstract from the peculiarities of the individual members of a set of items", or "disregard, in considering separate things, those characteristics which serve to distinguish them". In that event we are not left, as LIPSCHITZ maintains, with "the concept of the Number of the things considered"; what we get is rather a general concept under which the things in question fall. The things themselves do not in the process lose any of their special characteristics. If, for example, in considering a white cat and a black cat, I disregard the properties which serve to distinguish them, then I get presumably the concept "cat". Even if I proceed to bring them both under this concept and call them, I suppose, units, the white one still remains white just the same, and the black black. I may not think about their colours, or I may propose to make no inference from their difference in this respect, but for all that the cats do not become colourless and they remain different precisely as before. The concept "cat", no doubt, which we have arrived at by abstraction,

---

[1] Baumann, op. cit., Vol. I, p. 242 [op. cit., Dial. I, p. 16 = Molesworth edn., p. 18].

[2] Baumann, op. cit., Vol. II, p. 568 [*Enquiry concerning Human Understanding*, Sect. XII, part iii, § 131].

[3] Op. cit., p. 1.

* [*aequales*]

die Abstraction gewonnen ist, enthält zwar die Besonderheiten nicht mehr, ist aber eben dadurch nur Einer.

§ 35. Durch blos begriffliche Verfahrungsweisen gelingt
es nicht, verschiedene Dinge gleich zu machen; gelänge es
aber, so hätte man nicht mehr Dinge, sondern nur Ein Ding;
denn, wie Descartes*) sagt, die Zahl — besser: die Mehr-
zahl — in den Dingen entspringt aus deren Unterscheidung.
E. Schröder**) behauptet mit Recht: „Die Anforderung
Dinge zu zählen kann vernünftiger Weise nur gestellt werden,
wo solche Gegenstände vorliegen, welche deutlich von einander
unterscheidbar z. B. räumlich und zeitlich getrennt und gegen
einander abgegrenzt erscheinen." In der That erschwert
zuweilen die zu grosse Aehnlichkeit z. B. der Stäbe eines
Gitters die Zählung. Mit besonderer Schärfe drückt sich
W. Stanley Jevons***) in diesem Sinne aus: „Zahl ist
nur ein andrer Name für Verschiedenheit. Genaue Identität
ist Einheit, und mit Verschiedenheit entsteht Mehrheit."
Und weiter (S. 157): „Es ist oft gesagt, dass Einheiten
Einheiten sind, insofern sie einander volkommen gleichen;
aber, obwohl sie in einigen Rücksichten vollkommen gleich
sein mögen, müssen sie mindestens in Einem Punkte verschieden sein; sonst wäre der Begriff der Mehrheit auf sie
unanwendbar. Wenn drei Münzen so gleich wären, dass sie
denselben Raum zu derselben Zeit einnähmen, so wären sie
nicht drei Münzen, sondern Eine Münze."

§ 36. Aber es zeigt sich bald, dass die Ansicht von
der Verschiedenheit der Einheiten auf neue Schwierigkeiten
stösst. Jevons erklärt: „Eine Einheit (unit) ist irgendein
Gegenstand des Denkens, der von irgendeinem andern Gegenstande unterschieden werden kann, der als Einheit in derselben Aufgabe behandelt wird." Hier ist Einheit durch

*) Baumann a. a. O. Bd. I. S. 103.
**) A. a. O. S. 3.
***) The principles of Science, 3d. Ed. S. 156.

no longer includes the special characteristics of either, but of it, for just that reason, there is only one.

§ 35. We cannot succeed in making different things identical simply by dint of operations with concepts. But even if we did, we should then no longer have things in the plural, but only one thing; for, as DESCARTES[1] says, the number (or better, the plurality) in things arises from their diversity. And as E. SCHRÖDER[2] justly observes: "That things should be numbered is a reasonable demand only where the objects submitted appear clearly distinguishable from one another (for example, spatially and temporally separated) and isolated in contrast with one another." It does actually happen at times that too great similarity, for instance of the uprights in a railing, does make numbering difficult. W. S. JEVONS[3] makes this point with unusual force: "Number is but another name for *diversity*. Exact identity is unity, and with difference arises plurality." And again (p. 157)[*]: "It has often been said that units are units in respect of being perfectly similar to each other; but though they may be perfectly similar in some respects, they must be different in at least one point, otherwise they would be incapable of plurality. If three coins were so similar that they occupied the same place at the same time, they would not be three coins, but one."

§ 36. However, the view that units must be different comes up, as soon transpires, against fresh difficulties. JEVONS defines a unit as "any object of thought which can be discriminated from every other object treated as a unit in the same problem." But this is to define unit in terms of itself,

---

[1] Baumann, op. cit., Vol. I, p. 103 [*Principia*, Part I, § 60].
[2] Op. cit., p. 3.
[3] *The principles of science*, 3rd edn., p. 156 [1874 edn., p. 175].

[*] [1874 edn., p. 176]

sich selbst erklärt und der Zusatz „der von irgendeinem andern Gegenstande unterschieden werden kann" enthält keine nähere Bestimmung weil er selbstverständlich ist. Wir nennen den Gegenstand eben nur darum einen andern, weil wir ihn vom ersten unterscheiden können. Jevons*) sagt ferner: „Wenn ich das Symbol 5 schreibe, meine ich eigentlich

$$1 + 1 + 1 + 1 + 1$$

und es ist vollkommen klar, dass jede dieser Einheiten von jeder andern verschieden ist. Wenn erforderlich, kann ich sie so bezeichnen:

$$1' + 1'' + 1''' + 1'''' + 1'''''."$$

Gewiss ist es erforderlich, sie verschieden zu bezeichnen, wenn sie verschieden sind; sonst würde ja die grösste Verwirrung entstehen. Wenn schon die verschiedene Stelle, an der die Eins erschiene, eine Verschiedenheit bedeuten sollte, so müsste das als ausnahmslose Regel hingestellt werden, weil man sonst nie wüsste, ob $1 + 1$ 2 bedeuten solle oder 1. Dann müsste man die Gleichung $1 = 1$ verwerfen und wäre in der Verlegenheit, nie dasselbe Ding zum zweiten Male bezeichnen zu können. Das geht offenbar nicht an. Wenn man aber verschiedenen Dingen verschiedene Zeichen geben will, so ist nicht einzusehen, weshalb man in diesen noch einen gemeinsamen Bestandtheil festhält und nicht lieber statt

$$1' + 1'' + 1''' + 1'''' + 1'''''$$

schreibt

$$a + b + c + d + e.$$

Die Gleichheit ist doch nun einmal verloren gegangen, und die Andeutung einer gewissen Aehnlichkeit nützt nichts. So zerrinnt uns die Eins unter den Händen; wir behalten die Gegenstände mit allen ihren Besonderheiten. Diese Zeichen

$$1', 1'', 1'''$$

sind ein sprechender Ausdruck für die Verlegenheit: wir

*) A. a. O. S. 162.

and the qualifying clause "which can be discriminated from every other object" fails to describe it any more precisely, because it goes without saying: we call them other objects simply and solely because we can discriminate them from the first mentioned. JEVONS[1] goes on: "Whenever I use the symbol 5 I really mean

$$1 + 1 + 1 + 1 + 1,$$

and it is perfectly understood that each of these units is distinct from each other. If requisite I might mark them thus

$$1' + 1'' + 1''' + 1'''' + 1'''''. \text{ "}$$

Certainly it is requisite to mark them differently, if they are different: otherwise the utmost confusion must result. For if a difference simply in the position in which the 1 appears were to be made to mean of itself a difference in the unit, this convention would have to be laid down as a rule without any exception, or else we should never know whether $1 + 1$ was to be taken to mean 2 or 1. Accordingly, we should have to give up the equation $1 = 1$ and we should never, to our embarrassment, be able to mark the same thing twice. That obviously will not do. If, however, we adopt the alternative plan, of assigning different symbols to different things, it is hard to see why we still retain in our symbols a common element; why not write, instead of

$$1' + 1'' + 1''' + 1'''' + 1''''',$$

simply

$$a + b + c + d + e?$$

But now the identity of the units has been completely lost, and it helps not at all to point out that they are to some extent similar. So our one slips through our fingers; we are left with the objects in all their particularity. The symbols

$$1', \ 1'', \ 1'''$$

tell the tale of our embarrassment. We must have identity—

[1] Op. cit., p. 162 [1874 edn., p. 182].

haben die Gleichheit nöthig; deshalb die 1; wir haben die Verschiedenheit nöthig; deshalb die Indices, die nur leider die Gleichheit wieder aufheben.

§ 37. Bei andern Schriftstellern stossen wir auf dieselbe Schwierigkeit. Locke*) sagt: „Durch Wiederholung der Idee einer Einheit und Hinzufügung derselben zu einer andern Einheit machen wir demnach eine collective Idee, die durch das Wort „zwei" bezeichnet wird. Und wer das thun und so weitergehen kann, immer noch Eins hinzufügend zu der letzten collectiven Idee, die er von einer Zahl hatte, und ihr einen Namen geben kann, der kann zählen." Leibniz**) definirt Zahl als 1 und 1 und 1 oder als Einheiten. Hesse***) sagt: „Wenn man sich eine Vorstellung machen kann von der Einheit, die in der Algebra mit dem Zeichen 1 ausgedrückt wird, . . . so kann man sich auch eine zweite gleichberechtigte Einheit denken und weitere derselben Art. Die Vereinigung der zweiten mit der ersten zu einem Ganzen giebt die Zahl 2".

Hier ist auf die Beziehung zu achten, in der die Bedeutungen der Wörter „Einheit" und „Eins" zu einander stehen. Leibniz versteht unter Einheit einen Begriff, unter den die Eins und die Eins und die Eins fallen, wie er denn auch sagt: „Das Abstracte von Eins ist die Einheit." Locke und Hesse scheinen Einheit und Eins gleichbedeutend zu gebrauchen. Im Grunde thut dies wohl auch Leibniz; denn indem er die einzelnen Gegenstände, die unter den Begriff der Einheit fallen, sämmtlich Eins nennt, bezeichnet er mit diesem Worte nicht den einzelnen Gegenstand, sondern den Begriff, unter den sie fallen.

§ 38. Um nicht Verwirrung einreissen zu lassen, wird es jedoch gut sein, einen Unterschied zwischen Einheit

*) Baumann a. a. O. Bd. I. S. 409–411.
**) Baumann a. a. O. Bd. II. S. 3.
***) Vier Species. S. 2.

hence the 1; but we must have difference—hence the strokes; only unfortunately, the latter undo the work of the former.

§ 37. In other writers we meet with the same difficulty. In LOCKE[1] we read: "By the repeating ... of the idea of an unit, and joining it to another unit, we make thereof one collective idea, marked by the name two. And whosoever can do this and proceed on, still adding one more to the last collective idea which he had of any number, and give a name to it, may count." LEIBNIZ[2] defines number as 1 and 1 and 1 or as units. HESSE[3] writes: "Anyone who can form for himself an idea of the unit which in algebra is expressed by the symbol 1, ... can go on to conceive a second unit as good as the first, and then further units of the same sort. The union of the second with the first into a single whole yields the number 2."

In these passages, the relation between the meanings of the words "unit" and "one" should be noticed. LEIBNIZ understands by *unitas* a concept under which this one and that one and the other one fall, or as he also puts it: "Abstractum ab uno est *Unitas*.". LOCKE and HESSE seem to use unit and one to mean the same. Indeed LEIBNIZ, in the last analysis, does so too; for when he calls each individual object falling under his concept of *unitas* a *unum*, this word is being used to signify not the individual object but the concept under which they all fall.

§ 38. However, if confusion is not to become worse confounded, it is advisable to observe a strict distinction

---

[1] Baumann, op. cit., Vol. I, pp. 410–11 [*Essay*, Bk. II., cap. xvi, § 5].
[2] Baumann, op. cit., Vol. II, p. 3 [Erdmann edn., p. 53].
[3] *Vier Species* [Leipzig 1872], p. 2.

und Eins streng aufrecht zu erhalten. Man sagt „die Zahl Eins" und deutet mit dem bestimmten Artikel einen bestimmten, einzelnen Gegenstand der wissenschaftlichen Forschung an. Es giebt nicht verschiedene Zahlen Eins, sondern nur Eine. Wir haben in 1 einen Eigennamen, der als solcher eines Plurals ebenso unfähig ist wie „Friedrich der Grosse" oder „das chemische Element Gold." Es ist nicht Zufall und nicht eine ungenaue Bezeichnungsweise, dass man 1 ohne unterscheidende Striche schreibt. Die Gleichung

$$3 - 2 = 1$$

würde St. Jevons etwa so wiedergeben:

$$(1' + 1'' + 1''') - (1'' + 1''') = 1'$$

Was würde aber das Ergebniss von

$$(1' + 1'' + 1''') - (1'''' + 1''''')$$

sein? Jedenfalls nicht 1'. Daraus geht hervor, dass es nach seiner Auffassung nicht nur verschiedene Einsen, sondern auch verschiedene Zweien u. s. w. geben würde; denn $1'' + 1'''$ könnte nicht durch $1'''' + 1'''''$ vertreten werden. Man sieht hieraus recht deutlich, dass die Zahl nicht eine Anhäufung von Dingen ist. Die Arithmetik würde aufgehoben werden, wollte man statt der Eins, die immer dieselbe ist, verschiedene Dinge einführen, wenn auch in noch so ähnlichen Zeichen; gleich dürften sie ja ohne Fehler nicht sein. Man kann doch nicht annehmen, dass das tiefste Bedürfniss der Arithmetik eine fehlerhafte Schreibung sei. Darum ist es unmöglich 1 als Zeichen für verschiedene Gegenstände anzusehen, wie Island, Aldebaran, Solon u. dgl. Am greifbarsten wird der Unsinn, wenn man an den Fall denkt, dass eine Gleichung drei Wurzeln hat, nämlich 2 und 5 und 4. Schreibt man nun nach Jevons für 3:

$$1' + 1'' + 1''',$$

so würde 1' hier 2, 1'' 5 und 1''' 4 bedeuten, wenn man unter 1', 1'', 1''' Einheiten und folglich nach Jevons die hier vorliegenden Gegenstände des Denkens versteht. Wäre es dann nicht verständlicher für $1' + 1'' + 1'''$ zu schreiben

between unit and one. When we speak of "the number one", we indicate by means of the definite article a definite and unique object of scientific study. There are not divers numbers one, but only one. In 1 we have a proper name, which as such does not admit of a plural any more than "Frederick the Great" or "the chemical element gold". It is no accident, nor is it a notational inexactitude, that we write 1 without any strokes to mark differences. JEVONS would rewrite the equation

$$3 - 2 = 1$$

in some such way as this:

$$(1' + 1'' + 1''') - (1'' + 1''') = 1'.$$

But what would be the remainder of

$$(1' + 1'' + 1''') - (1'''' + 1''''')?$$

Certainly not 1'. It follows, therefore, that on his view there would be not only distinct ones but also distinct twos and so on; for 1'''' + 1''''' could not be substituted for 1'' + 1'''. This puts us in a position to see quite clearly that number is not an agglomeration of things. Arithmetic would come to a dead stop, if we tried to introduce in place of the number one, which is always the same, different distinct things, however similar the symbols for them; yet to make the symbols identical would be, of course, a mistake, and surely we cannot suppose that the mainspring of arithmetic is a piece of faulty notation. It is therefore impossible to regard 1 as a symbol for different distinct objects, for Iceland, Aldebaran, Solon, and so on. The absurdity can be best brought out by taking the case of an equation which has three roots, namely 2, 5 and 4. Suppose now with JEVONS we write for 3

$$1' + 1'' + 1''';$$

and let us take 1' and 1'' and 1''' to be units, that is, still following JEVONS, to be the objects currently under consideration. It follows that 1' would here mean 2, and 1'' 5, and 1''' 4. Then would it not be more intelligible, instead of 1' + 1'' + 1''', to write

50

$2 + 5 + 4$?

Ein Plural ist nur von Begriffswörtern möglich. Wenn man also von „Einheiten" spricht, so kann man dies Wort nicht gleichbedeutend mit dem Eigennamen „Eins" gebrauchen, sondern als Begriffswort. Wenn „Einheit" „zu zählender Gegenstand" bedeutet, so kann man nicht Zahl als Einheiten definiren. Wenn man unter „Einheit" einen Begriff versteht, der die Eins und nur diese unter sich fasst, so hat ein Plural keinen Sinn, und es ist wieder unmöglich, mit Leibniz Zahl als Einheiten oder als 1 und 1 und 1 zu definiren. Wenn man das „und" so gebraucht wie in „Bunsen und Kirchhof," so ist 1 und 1 und 1 nicht 3, sondern 1, sowie Gold und Gold und Gold nie etwas anderes als Gold ist. Das Pluszeichen in

$$1 + 1 + 1 = 3$$

muss also anders als das „und" aufgefasst werden, das eine Sammlung, eine „collective Idee" bezeichnen hilft.

§ 39. Wir stehen demnach vor folgender Schwierigkeit: Wenn wir die Zahl durch Zusammenfassung von verschiedenen Gegenständen entstehen lassen wollen, so erhalten wir eine Anhäufung, in der die Gegenstände mit eben den Eigenschaften enthalten sind, durch die sie sich unterscheiden, und das ist nicht die Zahl. Wenn wir die Zahl andrerseits durch Zusammenfassung von Gleichem bilden wollen, so fliesst dies immerfort in eins zusammen, und wir kommen nie zu einer Mehrheit.

Wenn wir mit 1 jeden der zu zählenden Gegenstände bezeichnen, so ist das ein Fehler, weil Verschiedenes dasselbe Zeichen erhält. Versehen wir die 1 mit unterscheidenden Strichen, so wird sie für die Arithmetik unbrauchbar.

Das Wort „Einheit" ist vortrefflich geeignet, diese Schwierigkeit zu verhüllen; und das ist der — wenn auch unbewusste — Grund, warum man es den Wörtern „Gegenstand" und „Ding" vorzieht. Man nennt zunächst die zu zählenden Dinge Einheiten, wobei die Verschiedenheit ihr

$$2 + 5 + 4?$$

Only concept words can form a plural. If, therefore, we speak of "units", we must be using the word not as equivalent to the proper name "one", but as a concept word. If this term "unit" means "object to be numbered", then number cannot be defined as units. But if we understand by "unit" a concept which includes under it the number one and nothing else, a plural makes no sense, and it becomes impossible once more to define number, with LEIBNIZ, as units, or as 1 and 1 and 1; for if "and" is used as in "Bunsen and Kirchhof", then 1 and 1 and 1 is not 3 but 1, just as gold and gold and gold is never anything else but gold. The plus symbol in

$$1 + 1 + 1 = 3$$

must, therefore, be interpreted differently from the "and" which we use in symbolizing a collection or a "collective idea".

§ 39. We are faced, therefore, with the following difficulty:

If we try to produce the number by putting together different distinct objects, the result is an agglomeration in which the objects contained remain still in possession of precisely those properties which serve to distinguish them from one another; and that is not the number. But if we try to do it in the other way, by putting together identicals, the result runs perpetually together into one and we never reach a plurality.

If we use 1 to stand for each of the objects to be numbered, we make the mistake of assigning the same symbol to different things. But if we provide the 1 with differentiating strokes, it becomes unusable for arithmetic.

The word "unit" is admirably adapted to conceal this difficulty; and that is the real, though no doubt unconscious, reason why we prefer it to the words "object" and "thing". We start by calling the things to be numbered "units", without detracting from their diversity; then subsequently the concept

51

Recht erhält; dann geht die Zusammenfassung, Sammlung, Vereinigung, Hinzufügung, oder wie man es sonst nennen will, in den Begriff der arithmetischen Addition über und das Begriffswort „Einheit" verwandelt sich unvermerkt in den Eigennamen „Eins". Damit hat man dann die Gleichheit. Wenn ich an den Buchstaben **u** ein **n** und daran ein **d** füge, so sieht jeder leicht ein, dass das nicht die Zahl 3 ist. Wenn ich aber **u, n** und **d** unter den Begriff „Einheit" bringe und nun für „**u** und **n** und **d**" sage „eine Einheit und eine Einheit und noch eine Einheit" oder „1 und 1 und 1", so glaubt man leicht damit die 3 zu haben. Die Schwierigkeit wird durch das Wort „Einheit" so gut versteckt, dass gewiss nur wenige Menschen eine Ahnung von ihr haben.

Hier könnte Mill mit Recht tadelnd von einem kunstfertigen Handhaben der Sprache reden; denn hier ist es nicht die äussere Erscheinung eines Denkvorganges, sondern es spiegelt einen solchen nur vor. Hier hat man in der That den Eindruck, als ob den von Gedanken leeren Worten eine gewisse geheimnissvolle Kraft beigelegt werde, wenn Verschiedenes blos dadurch, dass man es Einheit nennt, gleich werden soll.

Versuche, die Schwierigkeit zu überwinden.

§ 40. Wir betrachten nun einige Ausführungen, die sich als Versuche zur Ueberwindung dieser Schwierigkeit darstellen, wenn sie auch wohl nicht immer mit klarem Bewusstsein in dieser Absicht gemacht sind.

Man kann zunächst eine Eigenschaft des Raumes und der Zeit zu Hilfe rufen. Ein Raumpunkt ist nämlich von einem andern, eine Gerade oder Ebene von einer andern, congruente Körper, Flächen- oder Linienstücke von einander, für sich allein betrachtet, gar nicht zu unterscheiden, sondern nur in ihrem Zusammensein als Bestandtheile einer Gesammt-

of putting together (or collecting, or uniting, or annexing, or whatever we choose to call it) transforms itself into that of arithmetical addition, while the concept word "unit" changes unperceived into the proper name "one". And there we have our identity. If I annex to the letter *a* first an *n* and then a *d*, anyone can easily see that that is not the number 3. If, however, I bring the letters *a*, *n* and *d* under the concept "unit", and now, instead of "*a* and *n* and *d*", say "a unit and a unit and a further unit" or "1 and 1 and 1", we are quite prepared to believe that this does give us the number 3. The difficulty is so well hidden under the word "unit", that those who have any suspicion of its existence must surely be few at most.

Here, indeed, is an artful manipulation of language worthy of MILL's censure; for this is no outward manifestation of an inward process of thought, but only the illusion of one. Here we really do have the impression that words devoid of thought must possess some mysterious power, if what is different is to be made identical simply by being called a unit.

Attempts to overcome the difficulty.

§ 40. We consider next some detailed views which represent attempts to overcome this difficulty, although they have not always been produced with that end clearly and consciously in view.

The first suggestion is to call for assistance on a certain property of time and space, as follows. One point of space, considered by itself, is absolutely indistinguishable from another, and so is a straight line, or a plane, or one of a number of congruent bodies or areas or line-segments: they are distinguishable only when conjoined as elements in a single

anschauung. So scheint sich hier Gleichheit mit Unterscheidbarkeit zu vereinen. Aehnliches gilt von der Zeit. Daher meint wohl Hobbes,*) dass die Gleichheit der Einheiten anders als durch Theilung des Continuums entstehe, könne kaum gedacht werden. Thomae**) sagt: „Stellt man eine Menge von Individuen oder Einheiten im Raume vor und zählt man sie successive, wozu Zeit erforderlich ist, so bleibt bei aller Abstraction als unterscheidendes Merkmal der Einheiten noch ihre verschiedene Stellung im Raume und ihre verschiedene Aufeinanderfolge in der Zeit übrig."

Zunächst erhebt sich das Bedenken gegen eine solche Auffassungsweise, dass dann das Zählbare auf das Räumliche und Zeitliche beschränkt wäre. Schon Leibniz***) weist die Meinung der Scholastiker zurück, die Zahl entstehe aus der blossen Theilung des Continuums und könne nicht auf unkörperliche Dinge angewandt werden. Baumann†) betont die Unabhängigkeit von Zahl und Zeit. Der Begriff der Einheit sei auch ohne die Zeit denkbar. St. Jevons††) sagt: „Drei Münzen sind drei Münzen, ob wir sie nun nach einander zählen oder sie alle zugleich betrachten. In vielen Fällen ist weder Zeit noch Raum der Grund des Unterschiedes, sondern allein Qualität. Wir können Gewicht, Trägheit und Härte des Goldes als drei Eigenschaften auffassen, obgleich keine von diesen vor noch nach der andern ist weder im Raum noch in der Zeit. Jedes Mittel der Unterscheidung kann eine Quelle der Vielheit sein." Ich füge hinzu: wenn die gezählten Gegenstände nicht wirklich auf einander folgen, sondern nur nach einander gezählt werden, so kann die Zeit nicht der Grund der Unterscheidung sein. Denn, um sie nach einander zählen zu können, müssen wir schon

  *) Baumann a. a. O. Bd. I. S. 242.
 **) Elementare Theorie der analyt. Functionen, S. 1.
***) Baumann a. a. O. Bd. II. S. 2.
  †) A. a. O. Bd. II. S. 668.
 ††) The Principles of Science, S. 157.

total intuition. Here, therefore, we seem to get identity combined with distinguishability. With the parts of time, the same applies. This is presumably why HOBBES[1] holds it for hardly conceivable that the identity of units should result from anything but the division of the continuum. As THOMAE[2] puts it: "If we consider a set of individuals or units in space and number them one after the other, for which time is necessary, then, abstract as we will, there remain always as discriminating marks of the units their different positions in space and in the order of succession in time."

The first doubt that strikes us about any such view is that then nothing would be numerable except what is spatial and temporal. LEIBNIZ[3] long ago rebutted the view of the schoolmen that number results from the mere division of the continuum and cannot be applied to immaterial things. BAUMANN[4] dissociates number emphatically from time: he claims that the concept of the unit is thinkable even apart from time. JEVONS[5] writes: "Three coins are three coins, whether we count them successively or regard them all simultaneously. In many cases neither time nor space is the ground of difference, but pure quality alone enters. We can discriminate, for instance, the weight, inertia, and hardness of gold as three qualities, though none of these is before or after the other, either in space or time. Every means of discrimination may be a source of plurality." I would add that, if the objects numbered do not follow one after another in actual fact, but it is only that they are numbered one after another, then time cannot be the ground of discrimination between them. For, if we are to be able to number them one after another, we must already be in posses-

---

[1] Baumann, op. cit., Vol. I, p. 242 [loc. cit., p. 45^e above].
[2] Op. cit., p. 1.
[3] Baumann, op. cit., Vol. II, p. 2 [loc. cit., p. 31^e above].
[4] Op. cit., Vol. II, p. 668.
[5] Op. cit., p. 157 [1874 edn., p. 176].

unterscheidende Kennzeichen haben. Die Zeit ist nur ein psychologisches Erforderniss zum Zählen, hat aber mit dem Begriffe der Zahl nichts zu thun. Wenn man unräumliche und unzeitliche Gegenstände durch Raum- oder Zeitpunkte vertreten lässt, so kann dies vielleicht für die Ausführung der Zählung vortheilhaft sein; grundsätzlich wird aber dabei die Anwendbarkeit des Zahlbegriffes auf Unräumliches und Unzeitliches vorausgesetzt.

§ 41. Wird denn aber der Zweck der Vereinigung von Unterscheidbarkeit und Gleichheit wirklich erreicht, wenn wir von allen unterscheidenden Kennzeichen ausser den räumlichen und zeitlichen absehen? Nein! Wir sind der Lösung nicht um Einen Schritt näher gekommen. Die grössere oder geringere Aehnlichkeit der Gegenstände thut nichts zur Sache, wenn sie doch zuletzt aus einander gehalten werden müssen. Ich darf die einzelnen Punkte, Linien u. s. f. hier ebenso wenig alle mit i bezeichnen, als ich sie bei geometrischen Betrachtungen sämmtlich A nennen darf; denn hier wie dort ist es nöthig, sie zu unterscheiden. Nur für sich, ohne Rücksicht auf ihre räumlichen Beziehungen sind die Raumpunkte einander gleich. Soll ich sie aber zusammen-fassen, so muss ich sie in ihrem räumlichen Zusammensein betrachten, sonst schmelzen sie unrettbar in Einem zusammen. Punkte stellen in ihrer Gesammtheit vielleicht irgendeine sternbildartige Figur vor oder sind irgendwie auf einer Ge-raden angeordnet, gleiche Strecken bilden vielleicht mit den Endpunkten zusammenstossend eine einzige Strecke oder liegen getrennt von einander. Die so entstehenden Gebilde können für dieselbe Zahl ganz verschieden sein. So würden wir auch hier verschiedene Fünfen, Sechsen u. s. w. haben. Die Zeitpunkte sind durch kurze oder lange, gleiche oder ungleiche Zwischenzeiten getrennt. Alles dies sind Ver-hältnisse, die mit der Zahl an sich gar nichts zu thun haben. Ueberall mischt sich etwas Besonderes ein, worüber die Zahl in ihrer Allgemeinheit weit erhaben ist. Sogar ein

sion of distinguishing marks. Time is only a psychological necessity for numbering, it has nothing to do with the concept of number. We do represent objects which are non-spatial and non-temporal by spatial or temporal points, and this may perhaps be of advantage in carrying out the procedure of numbering; but it presupposes, fundamentally, that the concept of number is applicable to the non-spatial and the non-temporal.

§ 41. But further, supposing we do disregard all distinguishing marks except those of space and time, do we then really succeed in combining distinguishability with identity? Not at all. We are not one step nearer a solution. Whether the objects are so much more similar or so much less is beside the point, if they have still to be kept separate in the end. I cannot here symbolize the individual points, or lines or whatever it may be, all alike by 1, any more than for purposes of geometry I can call them one and all A; in the one case as in the other, it is essential to distinguish between them. It is only considered in themselves, and neglecting their spatial relations, that points of space are identical with one another; if I am to think of them together, I am bound then to consider them in their collocation in space, or else they fuse irretrievably together into one. Now points taken together as a group may perhaps fall into some pattern or other like a constellation or may equally arrange themselves somehow or other on a straight line; and a group of identical segments may lie perhaps with their end-points adjacent so as to combine into a single segment or perhaps at a distance from one another. Patterns produced in this way can be completely different while the number of their elements remains the same. So that here once again we should have different distinct fives, sixes, and so forth. Points of time, again, are separated by time intervals, long or short, equal or unequal. All these are relationships which have absolutely nothing to do with number as such. Pervading them all is an admixture of some special element, which number in its general form leaves far behind. Even a single moment itself has something *sui generis*, which serves to

einzelner Moment hat etwas Eigenthümliches, wodurch er sich etwa von einem Raumpunkte unterscheidet, und wovon nichts in dem Zahlbegriffe vorkommt.

§ 42. Auch der Ausweg, räumliche und zeitliche Anordnung durch einen allgemeinern Reihenbegriff zu ersetzen, führt nicht zum Ziele; denn die Stelle in der Reihe kann nicht der Grund des Unterscheidens der Gegenstände sein, weil diese schon irgendworan unterschieden sein müssen, um in eine Reihe geordnet werden zu können. Eine solche Anordnung setzt immer Beziehungen zwischen den Gegenständen voraus, seien es nun räumliche oder zeitliche oder logische oder Tonintervalle oder welche sonst, durch die man sich von einem zum andern leiten lässt, und die mit deren Unterscheidung nothwendig verbunden sind.

Wenn Hankel*) ein Object 1 mal, 2 mal, 3 mal denken oder setzen lässt, so scheint auch dies ein Versuch zu sein, die Unterscheidbarkeit mit der Gleichheit des zu Zählenden zu vereinen. Aber man sieht auch sofort, dass es kein gelungener ist; denn diese Vorstellungen oder Anschauungen desselben Gegenstandes müssen, um nicht in Eine zusammenzufliessen, irgendwie verschieden sein. Ich meine auch, dass man berechtigt ist, von 45 Millionen Deutschen zu sprechen, ohne vorher 45 Millionen mal einen Normal-Deutschen gedacht oder gesetzt zu haben; das möchte etwas umständlich sein.

§ 43. Wahrscheinlich um die Schwierigkeiten zu vermeiden, die sich ergeben, wenn man mit St. Jevons jedes Zeichen 1 einen der gezählten Gegenstände bedeuten lässt, will E. Schröder dadurch einen Gegenstand nur abbilden. Die Folge ist, dass er nur das Zahlzeichen, nicht die Zahl erklärt. Er sagt nämlich**): „Um nun ein Zeichen zu erhalten, welches fähig ist auszudrücken, wieviele jener

---

*) Theorie der complexen Zahlensysteme, S. 1.
**) Lehrbuch der Arithmetik und Algebra, S. 5 ff.

distinguish it from, say, a point of space, and of which there is no trace in the concept of number.

§ 42. Another way out is to invoke instead of spatial or temporal order a more generalized concept of series, but this too fails of its object; for their positions in the series cannot be the basis on which we distinguish the objects, since they must already have been distinguished somehow or other, for us to have been able to arrange them in a series. Any such arrangement always presupposes relations between the objects, whether spatial or temporal or logical relations, or relations of pitch or what not, which serve to lead us on from one object to the next and which are necessarily bound up with distinguishing between them.

When HANKEL[1] speaks of our thinking or putting a thing once or twice or three times, this too seems to be an attempt to combine in the things to be numbered distinguishability with identity. But it is obvious too at once that it is not successful; for his ideas or intuitions of the same object must, if they are not to coalesce into one, be different in some way or other. Moreover we are, I imagine, entitled to speak of 45 million Germans without having first to have thought or put an average German 45 million times, which might be somewhat tedious.

§ 43. It is probably in order to avoid the difficulties which JEVONS runs into through making each symbol 1 mean one of the objects numbered, that E. SCHRÖDER allows it only to copy the object. The consequence is that he gives a definition not of number but only of numerals. To give his own words[2]: "To arrive at a symbol capable of expressing *how many*

---

[1] Op. cit., p. 1.
[2] Op. cit., pp. 5 ff.

55

Einheiten*) vorhanden sind, richtet man die Aufmerksamkeit der Reihe nach einmal auf eine jede derselben und bildet sie mit einem Strich: 1 (eine Eins, ein Einer) ab; diese Einer setzt man in eine Zeile neben einander, verbindet sie jedoch unter sich durch das Zeichen + (plus), da sonst zum Beispiel 111 nach der gewöhnlichen Zahlenbezeichnung als einhundert und elf gelesen würde. Man erhält auf diese Weise ein Zeichen wie:

$$1 + 1 + 1 + 1 + 1,$$

dessen Zusammensetzung man dadurch beschreiben kann, dass man sagt:

„Eine natürliche Zahl ist eine Summe von Einern."

Hieraus sieht man, dass für Schröder die Zahl ein Zeichen ist. Was durch dies Zeichen ausgedrückt wird, das, was ich bisher Zahl genannt habe, setzt er mit den Worten „wieviele jener Einheiten vorhanden sind" als bekannt voraus. Auch unter dem Worte „Eins" versteht er das Zeichen 1, nicht dessen Bedeutung. Das Zeichen + dient ihm zunächst nur als äusserliches Verbindungsmittel ohne eignen Inhalt; erst später wird die Addition erklärt. Er hätte wohl kürzer so sagen können: man schreibt ebensoviele Zeichen 1 neben einander, als man zu zählende Gegenstände hat, und verbindet sie durch das Zeichen +. Die Null würde dadurch auszudrücken sein, dass man nichts hinschreibt.

§ 44. Um nicht die unterscheidenden Kennzeichen der Dinge in die Zahl mitaufzunehmen, sagt St. Jevons**):

„Es wird jetzt wenig schwierig sein, eine klare Vorstellung von der Zahlen-Abstraction zu bilden. Sie besteht im Abstrahiren von dem Charakter der Verschiedenheit, aus der Vielheit entspringt, indem man lediglich ihr Vorhandensein beibehält. Wenn ich von drei Männern spreche,

---

*) zu zählenden Gegenstände.
**) A. a. O. S. 158.

of such units[1] are present, we direct our attention upon each one of them in turn *once*, and copy it by a stroke (a *one*), thus, 1; these ones we put in a row side by side, only linking them up to each other by the symbol + (plus) because otherwise 111, for example, would be read, following the usual number notation, as one hundred and eleven. In this way we get a symbol such as:

$$1 + 1 + 1 + 1 + 1,$$

the composition of which we can describe by saying:

'A natural number is a sum of ones'."

This passage shows that for SCHRÖDER number is a *symbol*. What the symbol expresses, which is what I have been calling number, is taken, with the words "how many of such units are present", as already known. Even by the word "one" he understands the symbol 1, not its meaning. The symbol + is introduced solely to serve as a visible mark, without any content of its own, for linking up the other symbols; only later does he define addition. He could indeed have put what he means more briefly by saying that we write down side by side as many symbols 1 as we have objects to be numbered, and link them up by the symbol +. Nought would be expressed by writing down nothing.

§ 44. To avoid carrying over into number the distinguishing marks of the things numbered, JEVONS[2] invokes abstraction: "There will now be little difficulty in forming a clear notion of the nature of numerical abstraction. It consists in abstracting the character of the difference from which plurality arises, retaining merely the fact. When I speak of

---

[1] Objects to be numbered.
[2] Op. cit., p. 158 [1874 edn., p. 177].

so brauche ich nicht gleich die Kennzeichen einzeln anzugeben, an denen man jeden von ihnen von jedem unterscheiden kann. Diese Kennzeichen müssen vorhanden sein, wenn sie wirklich drei Männer und nicht ein und derselbe sind, und indem ich von ihnen als von vielen rede, behaupte ich damit zugleich das Vorhandensein der erforderlichen Unterschiede. Unbenannte Zahl ist also die leere Form der Verschiedenheit."

Wie ist das zu verstehn? Man kann entweder von den unterscheidenden Eigenschaften der Dinge abstrahiren, bevor man sie zu einem Ganzen vereinigt; oder man kann erst ein Ganzes bilden und dann von der Art der Unterschiede abstrahiren. Auf dem ersten Wege würden wir gar nicht zur Unterscheidung der Dinge kommen und also auch das Vorhandensein der Unterschiede nicht festhalten können; den zweiten Weg scheint Jevons zu meinen. Aber ich glaube nicht, dass wir so die Zahl 10000 gewinnen würden, weil wir nicht im Stande sind, so viele Unterschiede gleichzeitig aufzufassen und ihr Vorhandensein festzuhalten; denn, wenn es nach einander geschähe, so würde die Zahl nie fertig werden. Wir zählen zwar in der Zeit; aber dadurch gewinnen wir nicht die Zahl, sondern bestimmen sie nur. Uebrigens ist die Angabe der Weise des Abstrahirens keine Definition.

Was soll man sich unter der „leeren Form der Verschiedenheit" denken? etwa einen Satz wie

„a ist verschieden von b",

wobei a und b unbestimmt bleiben? Wäre dieser Satz etwa die Zahl 2? Ist der Satz

„die Erde hat zwei Pole"

gleichbedeutend mit

„der Nordpol ist vom Südpol verschieden"?

Offenbar nicht. Der zweite Satz könnte ohne den ersten und dieser ohne jenen bestehen. Für die Zahl 1000 würden wir dann

three men I need not at once specify the marks by which each may be known from each. Those marks must exist if they are really three men and not one and the same, and in speaking of them as many I imply the existence of the requisite differences. Abstract number, then, is *the empty form of difference*."

How are we to interpret this? Either we can abstract from the distinguishing properties of things before uniting them into a whole: or we can first form a whole and then abstract from the distinguishing properties. By the first method we should never get so far as to distinguish the things at all, and consequently could not retain the fact of the existence of the differences either; the second method seems to be what JEVONS intends. But by it we should never, it seems to me, arrive at a number like 10,000, for it is beyond our powers to grasp so many differences at once and retain the fact of their existence; while to go through them one after another is not enough, for the number would never be complete. We do our numbering in time, of course; but that does not give us the number itself, it only tells us the number of whatever it is we are numbering. Moreover, to tell us how to abstract is not, in any case, to give us a definition.

What are we to understand by "the empty form of difference"? Perhaps some proposition like

"*a* is different from *b*"

where *a* and *b* are left indefinite? Can this proposition be, say, the number 2? But does the proposition

"The Earth has two poles"

mean the same as

"The North Pole is different from the South"?

Obviously not. The second proposition could be true without the first being so, and vice versa. And for the number 1,000 we should then have as many as

solche Sätze haben, die eine Verschiedenheit ausdrücken.

Was Jevons sagt, passt insbesondere gar nicht auf die o und die 1. Wovon soll man eigentlich abstrahiren, um z. B. vom Monde auf die Zahl 1 zu kommen? Durch Abstrahiren erhält man wohl die Begriffe: Begleiter der Erde, Begleiter eines Planeten, Himmelskörper ohne eignes Licht, Himmelskörper, Körper, Gegenstand; aber die 1 ist in dieser Reihe nicht anzutreffen; denn sie ist kein Begriff, unter den der Mond fallen könnte. Bei der o hat man gar nicht einmal einen Gegenstand, von dem bei der Abstraction auszugehen wäre. Man wende nicht ein, dass o und 1 nicht Zahlen in demselben Sinne seien wie 2 und 3! Die Zahl antwortet auf die Frage wieviel? und wenn man z. B. fragt: wieviel Monde hat dieser Planet? so kann man sich ebenso gut auf die Antwort o oder 1 wie 2 oder 3 gefasst machen, ohne dass der Sinn der Frage ein andrer wird. Zwar hat die Zahl o etwas Besonderes und ebenso die 1, aber das gilt im Grunde von jeder ganzen Zahl; nur fällt es bei den grösseren immer weniger in die Augen. Es ist durchaus willkührlich, hier einen Artunterschied zu machen. Was nicht auf o oder 1 passt, kann für den Begriff der Zahl nicht wesentlich sein.

Endlich wird durch die Annahme dieser Entstehungsweise der Zahl die Schwierigkeit gar nicht gehoben, auf die wir bei der Betrachtung der Bezeichnung

$$1' + 1'' + 1''' + 1'''' + 1'''''$$

für 5 gestossen sind. Diese Schreibung steht gut im Einklange mit dem, was Jevons über die zahlenbildende Abstraction sagt; die obern Striche deuten nämlich an, dass eine Verschiedenheit da ist, ohne jedoch ihre Art anzugeben. Aber das blosse Bestehen der Verschiedenheit genügt schon, wie wir gesehen haben, um bei der Jevons'schen Auffassung verschiedene Einsen, Zweien, Dreien hervorzubringen,

$$\frac{1,000 \cdot 999}{1 \cdot 2}$$

such propositions, each stating a difference.

With the numbers 0 and 1 in particular, what JEVONS says simply will not work. What is it, in fact, that we are supposed to abstract from, in order to get, for example, from the moon to the number 1? By abstraction we do indeed get certain concepts, viz. satellite of the Earth, satellite of a planet, non-self-luminous heavenly body, heavenly body, body, object. But in this series 1 is not to be met with; for it is no concept that the moon could fall under. In the case of 0, we have simply no object at all from which to start our process of abstracting. It is no good objecting that 0 and 1 are not numbers in the same sense as 2 and 3. What answers the question How many? is number, and if we ask, for example, "How many moons has this planet?", we are quite as much prepared for the answer 0 or 1 as for 2 or 3, and that without having to understand the question differently. No doubt there is something unique about 0, and about 1 too; but the same is true in principle of every whole number, only the bigger the number the less obvious it is. To make out of this a difference in kind is utterly arbitrary. What will not work with 0 and 1 cannot be essential to the concept of number.

Finally, by taking number to arise in this manner we do not by any means remove the difficulty encountered when we were considering the symbolization of 5 by

$$1' + 1'' + 1''' + 1'''' + 1'''''.$$

This notation agrees well with what JEVONS says about the formation of number by abstraction; the strokes above the line, that is, indicate that a difference exists, without however specifying of what sort. But the mere existence of the difference is already enough, as we have seen, to produce on JEVONS' view different distinct ones and twos and threes,

was mit dem Bestande der Arithmetik durchaus unverträglich ist.

Lösung der Schwierigkeit.

§ 45. Ueberblicken wir nun das bisher von uns Festgestellte und die noch unbeantwortet gebliebenen Fragen!

Die Zahl ist nicht in der Weise wie Farbe, Gewicht, Härte von den Dingen abstrahirt, ist nicht in dem Sinne wie diese Eigenschaft der Dinge. Es blieb noch die Frage, von wem durch eine Zahlangabe etwas ausgesagt werde.

Die Zahl ist nichts Physikalisches, aber auch nichts Subjectives, keine Vorstellung.

Die Zahl entsteht nicht durch Hinzufügung von Ding zu Ding. Auch die Namengebung nach jeder Hinzufügung ändert darin nichts.

Die Ausdrücke „Vielheit," „Menge," „Mehrheit" sind wegen ihrer Unbestimmtheit ungeeignet, zur Erklärung der Zahl zu dienen.

In Bezug auf Eins und Einheit blieb die Frage, wie die Willkühr der Auffassung zu beschränken sei, die jeden Unterschied zwischen Einem und Vielen zu verwischen schien.

Die Abgegrenztheit, die Ungetheiltheit, die Unzerlegbarkeit sind keine brauchbaren Merkmale für das, was wir durch das Wort „Ein" ausdrücken.

Wenn man die zu zählenden Dinge Einheiten nennt, so ist die unbedingte Behauptung, dass die Einheiten gleich seien, falsch. Dass sie in gewisser Hinsicht gleich sind, ist zwar richtig aber werthlos. Die Verschiedenheit der zu zählenden Dinge ist sogar nothwendig, wenn die Zahl grösser als 1 werden soll.

So schien es, dass wir den Einheiten zwei widersprechende Eigenschaften beilegen müssten: die Gleichheit und die Unterscheidbarkeit.

Es ist ein Unterschied zwischen Eins und Einheit zu machen. Das Wort „Eins" ist als Eigenname eines Gegen-

which is utterly incompatible with the existence of arithmetic.

Solution of the difficulty.

§ 45. It is time now to survey what has been so far established and the questions which still remain unanswered.

Number is not abstracted from things in the way that colour, weight and hardness are, nor is it a property of things in the sense that they are. But when we make a statement of number*, what is that of which we assert something? This question remained unanswered.

Number is not anything physical, but nor is it anything subjective (an idea).

Number does not result from the annexing of thing to thing. It makes no difference even if we assign a fresh name after each act of annexation.

The terms "multitude", "set" and "plurality" are unsuitable, owing to their vagueness, for use in defining number.

In considering the terms one and unit, we left unanswered the question: How are we to curb the arbitrariness of our ways of regarding things, which threatens to obliterate every distinction between one and many?

Being isolated, being undivided, being incapable of dissection—none of these can serve as a criterion for what we express by the word "one".

If we call the things to be counted units, then the assertion that units are identical is, if made without qualification, false. That they are identical in this respect or that is true enough but of no interest. It is actually necessary that the things to be counted should be different if number is to get beyond 1.

We were thus forced, it seemed, to ascribe to units two contradictory qualities, namely identity and distinguishability.

A distinction must be drawn between one and unit. The word "one", as the proper name of an object of mathe-

* [See n. on p. 34e supra.]

standes der mathematischen Forschung eines Plurals unfähig. Es ist also sinnlos, Zahlen durch Zusammenfassen von Einsen entstehen zu lassen. Das Pluszeichen in $1 + 1 = 2$ kann nicht eine solche Zusammenfassung bedeuten.

§ 46. Um Licht in die Sache zu bringen, wird es gut sein, die Zahl im Zusammenhange eines Urtheils zu betrachten, wo ihre ursprüngliche Anwendungsweise hervortritt. Wenn ich in Ansehung derselben äussern Erscheinung mit derselben Wahrheit sagen kann: „dies ist eine Baumgruppe" und „dies sind fünf Bäume" oder „hier sind vier Compagnien" und „hier sind 500 Mann," so ändert sich dabei weder das Einzelne noch das Ganze, das Aggregat, sondern meine Benennung. Das ist aber nur das Zeichen der Ersetzung eines Begriffes durch einen andern. Damit wird uns als Antwort auf die erste Frage des vorigen Paragraphen nahe gelegt, dass die Zahlangabe eine Aussage von einem Begriffe enthalte. Am deutlichsten ist dies vielleicht bei der Zahl 0. Wenn ich sage: „die Venus hat 0 Monde", so ist gar kein Mond oder Aggregat von Monden da, von dem etwas ausgesagt werden könnte; aber dem Begriffe „Venusmond" wird dadurch eine Eigenschaft beigelegt, nämlich die, nichts unter sich zu befassen. Wenn ich sage: „der Wagen des Kaisers wird von vier Pferden gezogen," so lege ich die Zahl vier dem Begriffe „Pferd, das den Wagen des Kaisers zieht," bei.

Man mag einwenden, dass ein Begriff wie z. B. „Angehöriger des deutschen Reiches," obwohl seine Merkmale unverändert bleiben, eine von Jahr zu Jahr wechselnde Eigenschaft haben würde, wenn die Zahlangabe eine solche von ihm aussagte. Man kann dagegen geltend machen, dass auch Gegenstände ihre Eigenschaften ändern, was nicht verhindere, sie als dieselben anzuerkennen. Hier lässt sich aber der Grund noch genauer angeben. Der Begriff „Angehöriger des deutschen Reiches" enthält nämlich die Zeit als veränderlichen Bestandtheil, oder, um mich mathematisch

matical study, does not admit of a plural. Consequently, it is nonsense to make numbers result from the putting together of ones. The plus symbol in $1 + 1 = 2$ cannot mean such a putting together.

§ 46. It should throw some light on the matter to consider number in the context of a judgement which brings out its basic use. While looking at one and the same external phenomenon, I can say with equal truth both "It is a copse" and "It is five trees", or both "Here are four companies" and "Here are 500 men". Now what changes here from one judgement to the other is neither any individual object, nor the whole, the agglomeration of them, but rather my terminology. But that is itself only a sign that one concept has been substituted for another. This suggests as the answer to the first of the questions left open in our last paragraph, that the content of a statement of number is an assertion about a concept. This is perhaps clearest with the number 0. If I say "Venus has 0 moons", there simply does not exist any moon or agglomeration of moons for anything to be asserted of; but what happens is that a property is assigned to the *concept* "moon of Venus", namely that of including nothing under it. If I say "the King's carriage is drawn by four horses", then I assign the number four to the concept "horse that draws the King's carriage".

It may be objected that a concept like "inhabitant of Germany" would then possess, in spite of there being no change in its defining characteristics, a property which varied from year to year, if statements of the number of inhabitants did really assert a property of it. In reply to this, it is enough to point out that objects too can change their properties without that preventing us from recognizing them as the same. In this case, however, we can actually give the explanation more precisely. The fact is that the concept "inhabitant of Germany" contains a time-reference as a variable element in it, or, to put it mathematically, is a function of the time.

auszudrücken, ist eine Function der Zeit. Für „a ist ein Angehöriger des deutschen Reiches" kann man sagen: „a gehört dem deutschen Reiche an" und dies bezieht sich auf den gerade gegenwärtigen Zeitpunkt. So ist also in dem Begriffe selbst schon etwas Fliessendes. Dagegen kommt dem Begriffe „Angehöriger des deutschen Reiches zu Jahresanfang 1883 berliner Zeit" in alle Ewigkeit dieselbe Zahl zu.

§ 47. Dass eine Zahlangabe etwas Thatsächliches von unserer Auffassung Unabhängiges ausdrückt, kann nur den Wunder nehmen, welcher den Begriff für etwas Subjectives gleich der Vorstellung hält. Aber diese Ansicht ist falsch. Wenn wir z. B. den Begriff des Körpers dem des Schweren oder den des Wallfisches dem des Säugethiers unterordnen, so behaupten wir damit etwas Objectives. Wenn nun die Begriffe subjectiv wären, so wäre auch die Unterordnung des einen unter den andern als Beziehung zwischen ihnen etwas Subjectives wie eine Beziehung zwischen Vorstellungen. Freilich auf den ersten Blick scheint der Satz

„alle Wallfische sind Säugethiere"

von Thieren, nicht von Begriffen zu handeln; aber, wenn man fragt, von welchem Thiere denn die Rede sei, so kann man kein einziges aufweisen. Gesetzt, es liege ein Wallfisch vor, so behauptet doch von diesem unser Satz nichts. Man könnte aus ihm nicht schliessen, das vorliegende Thier sei ein Säugethier, ohne den Satz hinzuzunehmen, dass es ein Wallfisch ist, wovon unser Satz nichts enthält. Ueberhaupt ist es unmöglich, von einem Gegenstande zu sprechen, ohne ihn irgendwie zu bezeichnen oder zu benennen. Das Wort „Wallfisch" benennt aber kein Einzelwesen. Wenn man erwidert, allerdings sei nicht von einem einzelnen, bestimmten Gegenstande die Rede, wohl aber von einem unbestimmten, so meine ich, dass „unbestimmter Gegenstand" nur ein andrer Ausdruck für „Begriff" ist, und zwar ein schlechter, wider-

Instead of "*a* is an inhabitant of Germany" we can say "*a* inhabits Germany", and this refers to the current date at the time. Thus in the concept itself there is already something fluid. On the other hand, the number belonging to the concept "inhabitant of Germany at New Year 1883, Berlin time" is the same for all eternity.

§ 47. That a statement of number should express something factual independent of our way of regarding things can surprise only those who think a concept is something subjective like an idea. But this is a mistaken view. If, for example, we bring the concept of body under that of what has weight, or the concept of whale under that of mammal, we are asserting something objective; but if the concepts themselves were subjective, then the subordination of one to the other, being a relation between them, would be subjective too, just as a relation between ideas is. It is true that at first sight the proposition

"All whales are mammals"

seems to be not about concepts but about animals; but if we ask which animal then are we speaking of, we are unable to point to any one in particular. Even supposing a whale is before us, our proposition still does not state anything about it. We cannot infer from it that the animal before us is a mammal without the additional premiss that it is a whale, as to which our proposition says nothing. As a general principle, it is impossible to speak of an object without in some way designating or naming it; but the word "whale" is not the name of any individual creature. If it be replied that what we are speaking of is not, indeed, an individual definite object, but nevertheless an indefinite object, I suspect that "indefinite object" is only another term for concept, and a poor one at that, being

spruchsvoller. Mag immerhin unser Satz nur durch Beobachtung an einzelnen Thieren gerechtfertigt werden können, dies beweist nichts für seinen Inhalt. Für die Frage, wovon er handelt, ist es gleichgiltig, ob er wahr ist oder nicht, oder aus welchen Gründen wir ihn für wahr halten. Wenn nun der Begriff etwas Objectives ist, so kann auch eine Aussage von ihm etwas Thatsächliches enthalten.

§ 48. Der Schein, der vorhin bei einigen Beispielen entstand, dass demselben verschiedene Zahlen zukämen, erklärt sich daraus, dass dabei Gegenstände als Träger der Zahl angenommen wurden. Sobald wir den wahren Träger, den Begriff, in seine Rechte einsetzen, zeigen sich die Zahlen so ausschliessend wie in ihrem Bereiche die Farben.

Wir sehen nun auch, wie man dazu kommt, die Zahl durch Abstraction von den Dingen gewinnen zu wollen. Was man dadurch erhält, ist der Begriff, an dem man dann die Zahl entdeckt. So geht die Abstraction in der That oft der Bildung eines Zahlurtheils vorher. Die Verwechselung ist dieselbe, wie wenn man sagen wollte: der Begriff der Feuergefährlichkeit wird erhalten, indem man ein Wohnhaus aus Fachwerk mit einem Brettergiebel und Strohdach baut, dessen Schornsteine undicht sind.

Die sammelnde Kraft des Begriffes übertrifft weit die vereinigende der synthetischen Apperception. Durch diese wäre es nicht möglich, die Angehörigen des deutschen Reiches zu einem Ganzen zu verbinden; wohl aber kann man sie unter dem Begriff „Angehöriger des deutschen Reiches" bringen und zählen.

Nun wird auch die grosse Anwendbarkeit der Zahl erklärlich. Es ist in der That räthselhaft, wie dasselbe von äussern und zugleich von innern Erscheinungen, von Räumlichem und Zeitlichem und von Raum- und Zeitlosem ausgesagt werden könne. Dies findet nun in der Zahlangabe auch gar nicht statt. Nur den Begriffen, unter die das Aeussere

self-contradictory. However true it may be that our proposition can only be verified by observing particular animals, that proves nothing as to its content; to decide what it is about, we do not need to know whether it is true or not, nor for what reasons we believe it to be true. If, then, a concept is something objective, an assertion about a concept can have for its part a factual content.

§ 48. Several examples given earlier gave the false impression that different numbers may belong to the same thing. This is to be explained by the fact that we were there taking objects to be what has number. As soon as we restore possession to the rightful owner, the concept, numbers reveal themselves as no less mutually exclusive in their own sphere than colours are in theirs.

We now see also why there is a temptation to suggest that we get the number by abstraction from the things. What we do actually get by such means is the concept, and in this we then discover the number. Thus abstraction does genuinely often precede the formation of a judgement of number. It would be an analogous confusion to maintain that the way to acquire the concept of fire risk is to build a frame house, with timber gables, thatched roof and leaky chimneys.

The concept has a power of collecting together far superior to the unifying power of synthetic apperception. By means of the latter it would not be possible to join the inhabitants of Germany together into a whole; but we can certainly bring them all under the concept "inhabitant of Germany" and number them.

The wide range of applicability of number also now becomes explicable. Not without reason do we feel it puzzling that we should be able to assert the same predicate of physical and mental phenomena alike, of the spatial and temporal and of the non-spatial and non-temporal. But then, this simply is not what occurs with statements of number any more than elsewhere; numbers are assigned only to the concepts, under

und Innere, das Räumliche und Zeitliche, das Raum- und Zeitlose gebracht ist, werden Zahlen beigelegt.

§ 49. Wir finden für unsere Ansicht eine Bestätigung bei Spinoza, der sagt*): „Ich antworte, dass ein Ding blos rücksichtlich seiner Existenz, nicht aber seiner Essenz eines oder einzig genannt wird; denn wir stellen die Dinge unter Zahlen nur vor, nachdem sie auf ein gemeinsames Maass gebracht sind. Wer z. B. ein Sesterz und einen Imperial in der Hand hält, wird an die Zweizahl nicht denken, wenn er nicht dieses Sesterz und diesen Imperial mit einem und dem nämlichen Namen, nämlich Geldstück oder Münze belegen kann: dann kann er bejahen, dass er zwei Geldstücke oder Münzen habe; weil er nicht nur das Sesterz, sondern auch den Imperial mit dem Namen Münze bezeichnet," Wenn er fortfährt: „Hieraus ist klar, dass ein Ding eins oder einzig genannt wird, nur nachdem ein anderes Ding ist vorgestellt worden, das (wie gesagt) mit ihm übereinkommt," und wenn er meint, dass man nicht im eigentlichen Sinne Gott einen oder einzig nennen könne, weil wir von seiner Essenz keinen abstracten Begriff bilden könnten, so irrt er in der Meinung, der Begriff könne nur unmittelbar durch Abstraction von mehren Gegenständen gewonnen werden. Vielmehr kann man auch von den Merkmalen aus zu dem Begriffe gelangen; und dann ist es möglich, das kein Ding unter ihn fällt. Wenn dies nicht vorkäme, würde man nie die Existenz verneinen können, und damit verlöre auch die Behajung der Existenz ihren Inhalt.

§ 50. E. Schröder**) hebt hervor, dass, wenn von Häufigkeit eines Dinges solle gesprochen werden können, der Name dieses Dinges stets ein Gattungsname, ein allmeines Begriffswort (notio communis) sein müsse: „Sobald man nämlich einen Gegenstand vollständig — mit allen

*) Baumann a. a. O. Bd. I, S. 169.
**) A. a. O. S. 6.

which are brought both the physical and mental alike, both the spatial and temporal and the non-spatial and non-temporal.

§ 49. Corroboration for our view is to be found in SPINOZA,[1] where he writes: "I answer that a thing is called one or single simply with respect to its existence, and not with respect to its essence; for we only think of things in terms of number after they have first been reduced to a common genus. For example, a man who holds in his hand a sesterce and a dollar will not think of the number two unless he can cover his sesterce and his dollar with one and the same name, viz., piece of silver, or coin; then he can affirm that he has two pieces of silver, or two coins; since he designates by the name piece of silver or coin not only the sesterce but also the dollar." Unfortunately, he goes on: "From this it is clear, therefore, that nothing is called one or single except when some other thing has first been conceived which, as has been said, matches it", and he holds further that we cannot properly call God one or single, because it would be impossible for us to form an abstract concept of his essence. Here he makes the mistake of supposing that a concept can only be acquired by direct abstraction from a number of objects. We can, on the contrary, arrive at a concept equally well by starting from defining characteristics; and in such a case it is possible for nothing to fall under it. If this did not happen, we should never be able to deny existence, and so the assertion of existence too would lose all content.

§ 50. E. SCHRÖDER[2] calls attention to the fact that, if we are to be able to speak of the frequency of a thing, the name of the thing concerned must always be a *generic name*, a general concept word or *notio communis*; "So soon, that is, as we picture an object complete—with all its properties and in

---

[1] Baumann, op. cit., Vol. I, p. 169 [*Epistolae doctorum quorundam virorum*, No. 50 *ad* J. Jelles].
[2] Op. cit., p. 6.

seinen Eigenschaften und Beziehungen — in's Auge fasst, so wird derselbe einzig in der Welt dastehen und seines gleichen nicht weiter haben. Der Name des Gegenstandes wird alsdann den Charakter eines Eigennamens (nomen proprium) tragen und kann der Gegenstand nicht als ein wiederholt vorkommender gedacht werden. Dieses gilt aber nicht allein von concreten Gegenständen, es gilt überhaupt von jedem Dinge, mag dessen Vorstellung auch durch Abstractionen zu Stande kommen, wofern nur diese Vorstellung solche Elemente in sich schliesst, welche genügen, das betreffende Ding zu einem völlig bestimmten zu machen. . . . Das letztere" (Object der Zählung zu werden) „wird bei einem Dinge erst insofern möglich, als man von einigen ihm eigenthümlichen Merkmalen und Beziehungen, durch die es sich von allen andern Dingen unterscheidet, dabei absieht oder abstrahirt, wodurch dann erst der Name des Dinges zu einem auf mehre Dinge anwendbaren Begriffe wird."

§ 51. Das Wahre in dieser Ausführung ist in so schiefe und irreführende Ausdrücke gekleidet, dass eine Entwirrung und Sichtung geboten ist. Zunächst ist es unpassend, ein allgemeines Begriffswort Namen eines Dinges zu nennen. Dadurch entsteht der Schein als ob die Zahl Eigenschaft eines Dinges wäre. Ein allgemeines Begriffswort bezeichnet eben einen Begriff. Nur mit dem bestimmten Artikel oder einem Demonstrativpronomen gilt es als Eigenname eines Dinges, hört aber damit auf, als Begriffswort zu gelten. Der Name eines Dinges ist ein Eigenname. Ein Gegenstand kommt nicht wiederholt vor, sondern mehre Gegenstände fallen unter einen Begriff. Dass ein Begriff nicht nur durch Abstraction von den Dingen erhalten wird, die unter ihn fallen, ist schon Spinoza gegenüber bemerkt. Hier füge ich hinzu, dass ein Begriff dadurch nicht aufhört, Begriff zu sein, dass nur ein einziges Ding unter ihn fällt, welches demnach völlig durch ihn bestimmt ist. Einem solchen Begriffe (z. B. Begleiter der Erde) kommt eben die Zahl 1 zu,

all its relations, it will present itself as unique in the universe, and there will no longer be anything to match it. The name of the object takes on at once the character of a *proper name* (*nomen proprium*), and the object itself cannot be thought of as one which is found more than once. But observe that this holds good not only of *concrete* objects, but generally of anything and everything, even where the idea of it arises through *abstractions*, provided only that this idea contains in it sufficient elements to constitute the thing concerned a *completely* determinate thing. . .". For a thing to be numbered "first becomes possible in so far as, for that purpose, we disregard or *abstract from* some of its peculiar characteristics and relations, which distinguish it from all other things; this has the effect of turning what was the name of the thing into a concept applicable to more than one thing."

§ 51. What is true in this account is wrapped up in such distorted and misleading language, that we are obliged to straighten it out and sort the wheat from the chaff. To start with, it will not do to call a general concept word the name of a thing. That leads straight to the illusion that the number is a property of a thing. The business of a general concept word is precisely to signify a concept. Only when conjoined with the definite article or a demonstrative pronoun can it be counted as the proper name of a thing, but in that case it ceases to count as a concept word. The name of a thing is a proper name. An object, again, is not found more than once, but rather, more than one object falls under the same concept. That a concept need not be acquired by abstraction from the things which fall under it has already been pointed out in criticizing SPINOZA. Here I will add that a concept does not cease to be a concept simply because only one single thing falls under it, which thing, accordingly, is completely determined by it. It is to concepts of just this kind (for example, satellite of the Earth) that the number 1 belongs,

die in demselben Sinne Zahl ist wie 2 und 3. Bei einem Begriffe fragt es sich immer, ob etwas und was etwa unter ihn falle. Bei einem Eigennamen sind solche Fragen sinnlos. Man darf sich nicht dadurch täuschen lassen, dass die Sprache einen Eigennamen, z. B. Mond, als Begriffswort verwendet und umgekehrt; der Unterschied bleibt trotzdem bestehen. Sobald ein Wort mit dem unbestimmten Artikel oder im Plural ohne Artikel gebraucht wird, ist es Begriffswort.

§ 52. Eine weitere Bestätigung für die Ansicht, dass die Zahl Begriffen beigelegt wird, kann in dem deutschen Sprachgebrauche gefunden werden, dass man zehn Mann, vier Mark, drei Fass sagt. Der Singular mag hier andeuten, dass der Begriff gemeint ist, nicht das Ding. Der Vorzug dieser Ausdrucksweise tritt besonders bei der Zahl 0 hervor. Sonst freilich legt die Sprache den Gegenständen, nicht dem Begriffe Zahl bei: man sagt „Zahl der Ballen," wie man „Gewicht der Ballen" sagt. So spricht man scheinbar von Gegenständen, während man in Wahrheit von einem Begriffe etwas aussagen will. Dieser Sprachgebrauch ist verwirrend. Der Ausdruck „vier edle Rosse" erweckt den Schein, als ob „vier" den Begriff „edles Ross" ebenso wie „edel" den Begriff „Ross" näher bestimme. Jedoch ist nur „edel" ein solches Merkmal; durch das Wort „vier" sagen wir etwas von einem Begriffe aus.

§ 53. Unter Eigenschaften, die von einem Begriffe ausgesagt werden, verstehe ich natürlich nicht die Merkmale, die den Begriff zusammensetzen. Diese sind Eigenschaften der Dinge, die unter den Begriff fallen, nicht des Begriffes. So ist „rechtwinklig" nicht eine Eigenschaft des Begriffes „rechtwinkliges Dreieck"; aber der Satz, dass es kein rechtwinkliges, geradliniges, gleichseitiges Dreieck gebe, spricht eine Eigenschaft des Begriffes „rechtwinkliges, geradliniges, gleichseitiges Dreieck" aus; diesem wird die Nullzahl beigelegt.

which is a number in the same sense as 2 and 3. With a concept the question is always whether anything, and if so what, falls under it. With a proper name such questions make no sense. We should not be deceived by the fact that language makes use of proper names, for instance Moon, as concept words, and vice versa; this does not affect the distinction between the two. As soon as a word is used with the indefinite article or in the plural without any article, it is a concept word.

§ 52. Further confirmation of the view that number is assigned to concepts is to be found in idiom; just as in English we can speak of "three barrel", so in German we speak generally of "ten man", "four mark" and so on. The use of the singular here may indicate that the concept is intended, not the thing. The advantage of this way of speaking is particularly noticeable in the case of the number 0. Elsewhere, it must be admitted, our ordinary language does assign number not to concepts but to objects: we speak of "the number of the bales" just as we do of "the weight of the bales". Thus on the face of it we are talking about objects, whereas really we are intending to assert something of a concept. This usage is confusing. The construction in "four thoroughbred horses" fosters the illusion that "four" modifies the concept "thoroughbred horse" in just the same way as "thoroughbred" modifies the concept "horse." Whereas in fact only "thoroughbred" is a characteristic used in this way; the word "four" is used to assert something of a concept.

§ 53. By properties which are asserted of a concept I naturally do not mean the characteristics which make up the concept. These latter are properties of the things which fall under the concept, not of the concept. Thus "rectangular" is not a property of the concept "rectangular triangle"; but the proposition that there exists no rectangular equilateral rectilinear triangle does state a property of the concept "rectangular equilateral rectilinear triangle"; it assigns to it the number nought.

65

In dieser Beziehung hat die Existenz Aehnlichkeit mit der Zahl. Es ist ja Bejahung der Existenz nichts Anderes als Verneinung der Nullzahl. Weil Existenz Eigenschaft des Begriffes ist, erreicht der ontologische Beweis von der Existenz Gottes sein Ziel nicht. Ebensowenig wie die Existenz ist aber die Einzigkeit Merkmal des Begriffes „Gott". Die Einzigkeit kann nicht zur Definition dieses Begriffes gebraucht werden, wie man auch die Festigkeit, Geräumigkeit, Wohnlichkeit eines Hauses nicht mit Steinen, Mörtel und Balken zusammen bei seinem Baue verwenden kann. Man darf jedoch daraus, dass etwas Eigenschaft eines Begriffes ist, nicht allgemein schliessen, dass es aus dem Begriffe, d. h. aus dessen Merkmalen nicht gefolgert werden könne. Unter Umständen ist dies möglich, wie man aus der Art der Bausteine zuweilen einen Schluss auf die Dauerhaftigkeit eines Gebäudes machen kann. Daher wäre es zuviel behauptet, dass niemals aus den Merkmalen eines Begriffes auf die Einzigkeit oder Existenz geschlossen werden könne; nur kann dies nie so unmittelbar geschehen, wie man das Merkmal eines Begriffes einem unter ihn fallenden Gegenstande als Eigenschaft beilegt.

Es wäre auch falsch zu leugnen, dass Existenz und Einzigkeit jemals Merkmale von Begriffen sein könnten. Sie sind nur nicht Merkmale der Begriffe, denen man sie der Sprache folgend zuschreiben möchte. Wenn man z. B. alle Begriffe, unter welche nur Ein Gegenstand fällt, unter einen Begriff sammelt, so ist die Einzigkeit Merkmal dieses Begriffes. Unter ihn würde z. B. der Begriff „Erdmond," aber nicht der sogenannte Himmelskörper fallen. So kann man einen Begriff unter einen höhern, so zu sagen einen Begriff zweiter Ordnung fallen lassen. Dies Verhältniss ist aber nicht mit dem der Unterordnung zu verwechseln.

§ 54. Jetzt wird es möglich sein, die Einheit befriedigend zu erklären. E. Schröder sagt auf S. 7 seines genannten Lehrbuches: „Jener Gattungsname oder Begriff

In this respect existence is analogous to number. Affirmation of existence is in fact nothing but denial of the number nought. Because existence is a property of concepts the ontological argument for the existence of God breaks down. But oneness* is not a component characteristic of the concept "God" any more than existence is. Oneness cannot be used in the definition of this concept any more than the solidity of a house, or its commodiousness or desirability, can be used in building it along with the beams, bricks and mortar. However, it would be wrong to conclude that it is in principle impossible ever to deduce from a concept, that is, from its component characteristics, anything which is a property of the concept. Under certain conditions this is possible, just as we can occasionally infer the durability of a building from the type of stone used in building it. It would therefore be going too far to assert that we can never infer from the component characteristics of a concept to oneness or to existence; what is true is, that this can never be so direct a matter as it is to assign some component of a concept as a property to an object falling under it.

It would also be wrong to deny that existence and oneness can ever themselves be component characteristics of a concept. What is true is only that they are not components of those particular concepts to which language might tempt us to ascribe them. If, for example, we collect under a single concept all concepts under which there falls only one object, then oneness is a component characteristic of this new concept. Under it would fall, for example, the concept "moon of the Earth", though not the actual heavenly body called by this name. In this way we can make one concept fall under another higher or, so to say, second order concept. This relationship, however, should not be confused with the subordination of species to genus.

§ 54. It now becomes possible to give a satisfactory definition of the term "unit". E. SCHRÖDER writes, on p. 7 of his text book already referred to: "This generic name or

* [I.e. the character of being single or unique, called by theologians "unity".]

wird die Benennung der auf die angegebene Weise gebildeten Zahl genannt und macht das Wesen ihrer Einheit aus."

In der That, wäre es nicht am passendsten, einen Begriff Einheit zu nennen in Bezug auf die Anzahl, welche ihm zukommt? Wir können dann den Aussagen über die Einheit, dass sie von der Umgebung abgesondert und untheilbar sei, einen Sinn abgewinnen. Denn der Begriff, dem die Zahl beigelegt wird, grenzt im Allgemeinen das unter ihn Fallende in bestimmter Weise ab. Der Begriff „Buchstabe des Wortes Zahl" grenzt das Z gegen das a, dieses gegen das h u. s. w. ab. Der Begriff „Silbe des Wortes Zahl" hebt das Wort als ein Ganzes und in dem Sinne Untheilbares heraus, dass die Theile nicht mehr unter den Begriff „Silbe des Wortes Zahl" fallen. Nicht alle Begriffe sind so beschaffen. Wir können z. B. das unter den Begriff des Rothen Fallende in mannigfacher Weise zertheilen, ohne dass die Theile aufhören, unter ihn zu fallen. Einem solchen Begriffe kommt keine endliche Zahl zu. Der Satz von der Abgegrenztheit und Untheilbarkeit der Einheit lässt sich demnach so aussprechen:

Einheit in Bezug auf eine endliche Anzahl kann nur ein solcher Begriff sein, der das unter ihn Fallende bestimmt abgrenzt und keine beliebige Zertheilung gestattet.

Man sieht aber, dass Untheilbarkeit hier eine besondere Bedeutung hat.

Nun beantworten wir leicht die Frage, wie die Gleichheit mit der Unterscheidbarkeit der Einheiten zu versöhnen sei. Das Wort „Einheit" ist hier in doppeltem Sinne gebraucht. Gleich sind die Einheiten in der oben erklärten Bedeutung dieses Worts. In dem Satze: „Jupiter hat vier Monde" ist die Einheit „Jupitersmond". Unter diesen Begriff fällt sowohl I als auch II, als auch III, als auch IV. Daher kann man sagen: die Einheit, auf die I bezogen wird, ist gleich der Einheit, auf die II bezogen wird u. s. f. Da haben wir die Gleichheit. Wenn man aber die Unterscheid-

concept will be called the denomination of the number formed by the method given, and constitutes, in effect, what is meant by its unit."

Why not, in fact, adopt this very apt suggestion, and call a concept the unit relative to the Number which belongs to it? We can then achieve a sense for the assertions made about the unit, that it is isolated from its environment and is indivisible. For it is the case that the concept, to which the number is assigned, does in general isolate in a definite manner what falls under it. The concept "letters in the word three" isolates the *t* from the *h*, the *h* from the *r*, and so on. The concept "syllables in the word three" picks out the word as a whole, and as indivisible in the sense that no part of it falls any longer under that same concept. Not all concepts possess this quality. We can, for example, divide up something falling under the concept "red" into parts in a variety of ways, without the parts thereby ceasing to fall under the same concept "red". To a concept of this kind no finite number will belong. The proposition asserting that units are isolated and indivisible can, accordingly, be formulated as follows:

Only a concept which isolates what falls under it in a definite manner, and which does not permit any arbitrary division of it into parts, can be a unit relative to a finite Number.

It will be noticed, however, that "indivisibility" here has a special meaning.

We can now easily solve the problem of reconciling the identity of units with their distinguishability. The word "unit" is being used here in a double sense. The units are identical if the word has the meaning just explained. In the proposition "Jupiter has four moons", the unit is "moon of Jupiter". Under this concept falls moon I, and likewise also moon II, and moon III too, and finally moon IV. Thus we can say: the unit to which I relates is identical with the unit to which II relates, and so on. This gives us our identity.

barkeit der Einheiten behauptet, so versteht man darunter die der gezählten Dinge.

IV. Der Begriff der Anzahl.

Jede einzelne Zahl ist ein selbständiger Gegenstand.

§ 55. Nachdem wir erkannt haben, dass die Zahlangabe eine Aussage von einem Begriffe enthält, können wir versuchen, die leibnizischen Definitionen der einzelnen Zahlen durch die der o und der 1 zu ergänzen.

Es liegt nahe zu erklären: einem Begriffe kommt die Zahl o zu, wenn kein Gegenstand unter ihn fällt. Aber hier scheint an die Stelle der o das gleichbedeutende „kein" getreten zu sein; deshalb ist folgender Wortlaut vorzuziehen: einem Begriffe kommt die Zahl o zu, wenn allgemein, was auch a sei, der Satz gilt, dass a nicht unter diesen Begriff falle.

In ähnlicher Weise könnte man sagen: einem Begriffe F kommt die Zahl 1 zu, wenn nicht allgemein, was auch a sei, der Satz gilt, dass a nicht unter F falle, und wenn aus den Sätzen

„a fällt unter F" und „b fällt unter F"

allgemein folgt, dass a und b dasselbe sind.

Es bleibt noch übrig, den Uebergang von einer Zahl zur nächstfolgenden allgemein zu erklären. Wir versuchen folgenden Wortlaut: dem Begriffe F kommt die Zahl $(n+1)$ zu, wenn es einen Gegenstand a giebt, der unter F fällt und so beschaffen ist, dass dem Begriffe „unter F fallend, aber nicht a" die Zahl n zukommt.

§ 56. Diese Erklärungen bieten sich nach unsern bisherigen Ergebnissen so ungezwungen dar, dass es einer Darlegung bedarf, warum sie uns nicht genügen können.

Am ehesten wird die letzte Definition Bedenken erregen; denn genau genommen ist uns der Sinn des Ausdruckes

But when we assert the distinguishability of units, we mean that the things numbered are distinguishable.

IV. The concept of Number.

Every individual number is a self-subsistent object.

§ 55. Now that we have learned that the content of a statement of number is an assertion about a concept, we can try to complete the Leibnizian definitions of the individual numbers by giving the definitions of o and of 1.

It is tempting to define o by saying that the number o belongs to a concept if no object falls under it. But this seems to amount to replacing o by "no", which means the same. The following formulation is therefore preferable: the number o belongs to a concept, if the proposition that *a* does not fall under that concept is true universally, whatever *a* may be.

Similarly we could say: the number 1 belongs to a concept *F*, if the proposition that *a* does not fall under *F* is not true universally, whatever *a* may be, and if from the propositions

"*a* falls under *F*" and "*b* falls under *F*"

it follows universally that *a* and *b* are the same.

It remains still to give a general definition of the step from any given number to the next. Let us try the following formulation: the number $(n + 1)$ belongs to a concept *F*, if there is an object *a* falling under *F* and such that the number *n* belongs to the concept "falling under *F*, but not *a*".

§ 56. These definitions suggest themselves so spontaneously in the light of our previous results, that we shall have to go into the reasons why they cannot be reckoned satisfactory.

The most likely to cause misgivings is the last; for strictly speaking we do not know the sense of the expression "the

„dem Begriffe G kommt die Zahl n zu" ebenso unbekannt
wie der des Ausdruckes „dem Begriffe F kommt die Zahl
(n + 1) zu." Zwar können wir mittels dieser und der vor-
letzten Erklärung sagen, was es bedeute
„dem Begriffe F kommt die Zahl 1 + 1 zu,"
und dann, indem wir dies benutzen, den Sinn des Ausdruckes
„dem Begriffe F kommt die Zahl 1 + 1 + 1 zu"
angeben u. s. w.; aber wir können — um ein krasses Beispiel
zu geben — durch unsere Definitionen nie entscheiden, ob
einem Begriffe die Zahl Julius Caesar zukomme, ob dieser
bekannte Eroberer Galliens eine Zahl ist oder nicht. Wir
können ferner mit Hilfe unserer Erklärungsversuche nicht
beweisen, dass a = b sein muss, wenn dem Begriffe F die
Zahl a zukommt, und wenn demselben die Zahl b zukommt.
Der Ausdruck „die Zahl, welche dem Begriffe F zukommt"
wäre also nicht zu rechtfertigen und dadurch würde es überhaupt unmöglich, eine Zahlengleichheit zu beweisen, weil
wir gar nicht eine bestimmte Zahl fassen könnten. Es ist
nur Schein, dass wir die 0, die 1 erklärt haben; in Wahrheit haben wir nur den Sinn der Redensarten
„die Zahl 0 kommt zu,"
„die Zahl 1 kommt zu"
festgestellt; aber es nicht erlaubt, hierin die 0, die 1 als
selbständige, wiedererkennbare Gegenstände zu unterscheiden.

§ 57. Es ist hier der Ort, unsern Ausdruck, dass die
Zahlangabe eine Aussage von einem Begriffe enthalte, etwas
genauer ins Auge zu fassen. In dem Satze „dem Begriffe F
kommt die Zahl 0 zu" ist 0 nur ein Theil des Praedicates,
wenn wir als sachliches Subject den Begriff F betrachten.
Deshalb habe ich es vermieden, eine Zahl wie 0, 1, 2
Eigenschaft eines Begriffes zu nennen. Die einzelne Zahl
erscheint eben dadurch dass sie nur einen Theil der Aussage
bildet, als selbständiger Gegenstand. Ich habe schon oben
darauf aufmerksam gemacht, dass man „die 1" sagt und
durch den bestimmten Artikel 1 als Gegenstand hinstellt.

number *n* belongs to the concept *G*" any more than we do that of the expression "the number $(n + 1)$ belongs to the concept *F*". We can, of course, by using the last two definitions together, say what is meant by

"the number $1 + 1$ belongs to the concept *F*"

and then, using this, give the sense of the expression

"the number $1 + 1 + 1$ belongs to the concept *F*"

and so on; but we can never—to take a crude example—decide by means of our definitions whether any concept has the number JULIUS CAESAR belonging to it, or whether that same familiar conqueror of Gaul is a number or is not. Moreover we cannot by the aid of our suggested definitions prove that, if the number *a* belongs to the concept *F* and the number *b* belongs to the same concept, then necessarily $a = b$. Thus we should be unable to justify the expression "*the* number which belongs to the concept *F*", and therefore should find it impossible in general to prove a numerical identity, since we should be quite unable to achieve a determinate number. It is only an illusion that we have defined 0 and 1; in reality we have only fixed the sense of the phrases

"the number 0 belongs to"

"the number 1 belongs to";

but we have no authority to pick out the 0 and 1 here as self-subsistent objects that can be recognized as the same again.

§ 57. It is time to get a clearer view of what we mean by our expression "the content of a statement of number is an assertion about a concept". In the proposition "the number 0 belongs to the concept *F*", 0 is only an element in the predicate (taking the concept *F* to be the real subject). For this reason I have avoided calling a number such as 0 or 1 or 2 a *property* of a concept. Precisely because it forms only an element in what is asserted, the individual number shows itself for what it is, a self-subsistent object. I have already drawn attention above to the fact that we speak of "the number 1", where the definite article serves to class it as an object. In arithmetic this self-

Diese Selbständigkeit zeigt sich überall in der Arithmetik, z. B. in der Gleichung $1 + 1 = 2$. Da es uns hier darauf ankommt, den Zahlbegriff so zu fassen, wie er für die Wissenschaft brauchbar ist, so darf es uns nicht stören, dass im Sprachgebrauche des Lebens die Zahl auch attributiv erscheint. Dies lässt sich immer vermeiden. Z.B. kann man den Satz „Jupiter hat vier Monde" umsetzen in „die Zahl der Jupitersmonde ist vier.". Hier darf das „ist" nicht als blosse Copula betrachtet werden, wie in dem Satze „der Himmel ist blau". Das zeigt sich darin, dass man sagen kann: „die Zahl der Jupitersmonde ist die vier" oder „ist die Zahl 4". Hier hat „ist" den Sinn von „ist gleich," „ist dasselbe wie". Wir haben also eine Gleichung, die behauptet, dass der Ausdruck "die Zahl der Jupitersmonde" denselben Gegenstand bezeichne wie das Wort „vier". Und die Form der Gleichung ist die herrschende in der Arithmetik. Gegen diese Auffassung streitet nicht, dass in dem Worte „vier" nichts von Jupiter oder von Mond enthalten ist. Auch in dem Namen „Columbus" liegt nichts von Entdecken oder von Amerika und dennoch wird derselbe Mann Columbus und der Entdecker Amerikas genannt.

§ 58. Man könnte einwenden, dass wir uns von dem Gegenstande, den wir Vier oder die Anzahl der Jupitersmonde nennen, als von etwas Selbständigem durchaus keine Vorstellung*) machen können. Aber die Selbständigkeit, die wir der Zahl gegeben haben, ist nicht Schuld daran. Zwar glaubt man leicht, dass in der Vorstellung von vier Augen eines Würfels etwas vorkomme, was dem Worte „vier" entspräche; aber das ist Täuschung. Man denke an eine grüne Wiese und versuche, ob sich die Vorstellung ändert, wenn man den unbestimmten Artikel durch das Zahlwort „Ein" ersetzt. Es kommt nichts hinzu, während doch dem Worte „grün" etwas in der Vorstellung entspricht.

---

*) „Vorstellung" in dem Sinne von etwas Bildartigem genommen.

subsistence comes out at every turn, as for example in the identity $1 + 1 = 2$. Now our concern here is to arrive at a concept of number usable for the purposes of science; we should not, therefore, be deterred by the fact that in the language of everyday life number appears also in attributive constructions. That can always be got round. For example, the proposition "Jupiter has four moons" can be converted into "the number of Jupiter's moons is four". Here the word "is" should not be taken as a mere copula, as in the proposition "the sky is blue". This is shown by the fact that we can say: "the number of Jupiter's moons is the number four, or 4" Here "is" has the sense of "is identical with" or "is the same as". So that what we have is an identity, stating that the expression "the number of Jupiter's moons" signifies the same object as the word "four". And identities are, of all forms of proposition, the most typical of arithmetic. It is no objection to this account that the word "four" contains nothing about Jupiter or moons. No more is there in the name "Columbus" anything about discovery or about America, yet for all that it is the same man that we call Columbus and the discoverer of America.

§ 58. A possible criticism is, that we are not able to form of this object which we are calling Four or the Number of Jupiter's moons any sort of idea[1] at all which would make it something self-subsistent. But that is not the fault of the self-subsistence we have ascribed to the number. It is easy, I know, to suppose that in our idea of four dots on a die there is to be found something which corresponds to the word "four"; but that is a misapprehension. We have only to think of a green field, and try whether the idea alters when we replace the indefinite article by the number word "one"; nothing fresh is added—whereas with the word "green", there really is in the idea something which corresponds to it. If we

---

[1] "Idea" in the sense of something like a picture.

70

Wenn man sich das gedruckte Wort „Gold" vorstellt, wird man zunächst an keine Zahl dabei denken. Fragt man sich nun, aus wieviel Buchstaben es bestehe, so ergiebt sich die Zahl 4; aber die Vorstellung wird dadurch nicht etwa bestimmter, sondern kann ganz unverändert bleiben. Der hinzutretende Begriff „Buchstabe des Wortes Gold" ist eben das, woran wir die Zahl entdecken. Bei den vier Augen eines Würfels ist die Sache etwas versteckter, weil der Begriff sich uns durch die Aehnlichkeit der Augen so unmittelbar aufdrängt, dass wir sein Dazwischentreten kaum bemerken. Die Zahl kann weder als selbständiger Gegenstand noch als Eigenschaft an einem äussern Dinge vorgestellt werden, weil sie weder etwas Sinnliches noch Eigenschaft eines äussern Dinges ist. Am deutlichsten ist die Sache wohl bei der Zahl o. Man wird vergebens versuchen, sich o sichtbare Sterne vorzustellen. Zwar kann man sich den Himmel ganz mit Wolken überzogen denken; aber hierin ist nichts, was dem Worte „Stern" oder der o entspräche. Man stellt sich nur eine Sachlage vor, die zu dem Urtheile veranlassen kann: es ist jetzt kein Stern zu sehen.

§ 59. Jedes Wort erweckt vielleicht irgendeine Vorstellung in uns, sogar ein solches wie „nur"; aber sie braucht nicht dem Inhalte des Wortes zu entsprechen; sie kann in andern Menschen eine ganz andere sein. Man wird sich dann wohl eine Sachlage vorstellen, die zu einem Satze auffordert, in welchem das Wort vorkommt; oder es ruft etwa das gesprochene Wort das geschriebene ins Gedächtniss zurück.

Dies findet nicht nur bei Partikeln statt. Es unterliegt wohl keinem Zweifel, dass wir keine Vorstellung unserer Entfernung von der Sonne haben. Denn, wenn wir auch die Regel kennen, wie oft wir einen Maasstab vervielfältigen müssen, so misslingt doch jeder Versuch, nach dieser Regel uns ein Bild zu entwerfen, das auch nur einigermaassen dem Gewollten nahe kommt. Das ist aber kein Grund, die Rich-

imagine the printed word "gold", we shall not immediately think of any number in connexion with it. If we now ask ourselves how many letters it contains, the number 4 is the result; yet the idea does not become in consequence any more definite, but may remain completely unaltered. Where we discover the number is precisely in the freshly added concept "letter in the word gold". In the case of the four dots on the die, the matter is rather more obscured, because the concept thrusts itself upon us so immediately, owing to the similarity of the dots, that we scarcely notice its intervention. We can form no idea of the number either as a self-subsistent object or as a property in an external thing, because it is not in fact either anything sensible or a property of an external thing. But the point is clearest in the case of the number o; we shall try in vain to form an idea of o visible stars. We can, of course, think of a sky entirely overcast with clouds; but in this there is nothing to correspond to the word "star" or to o. All we succeed in imagining is a situation where the natural judgement to make would be: No star is now to be seen.

§ 59. It may be that every word calls up some sort of idea in us, even a word like "only"; but this idea need not correspond to the content of the word; it may be quite different in different men. The sort of thing we do is to imagine a situation where some proposition in which the word occurs would be called for; or it may happen that the spoken word recalls the written word to our memory.

Nor does this happen only in the case of particles. There is not the slightest doubt that we can form no idea of our distance from the sun. For even although we know the rule that we must multiply a measuring rod so and so many times, we still fail in every attempt to construct by its means a picture approximating even faintly to what we want. Yet this is no

tigkeit der Rechnung zu bezweifeln, durch welche die Entfernung gefunden ist, und hindert uns in keiner Weise, weitere Schlüsse auf das Bestehen dieser Entfernung zu gründen.

§ 60. Selbst ein so concretes Ding wie die Erde können wir uns nicht so vorstellen, wie wir erkannt haben, dass es ist; sondern wir begnügen uns mit einer Kugel von mässiger Grösse, die uns als Zeichen für die Erde gilt; aber wir wissen, dass diese sehr davon verschieden ist. Obwohl nun unsere Vorstellung das Gewollte oft gar nicht trifft, so urtheilen wir doch mit grosser Sicherheit über einen Gegenstand wie die Erde auch da, wo die Grösse in Betracht kommt.

Wir werden durch das Denken gar oft über das Vorstellbare hinausgeführt, ohne damit die Unterlage für unsere Schlüsse zu verlieren. Wenn auch, wie es scheint, uns Menschen Denken ohne Vorstellungen unmöglich ist, so kann doch deren Zusammenhang mit dem Gedachten ganz äusserlich, willkührlich und conventionell sein.

Es ist also die Unvorstellbarkeit des Inhaltes eines Wortes kein Grund, ihm jede Bedeutung abzusprechen oder es vom Gebrauche auszuschliessen. Der Schein des Gegentheils entsteht wohl dadurch, dass wir die Wörter vereinzelt betrachten und nach ihrer Bedeutung fragen, für welche wir dann eine Vorstellung nehmen. So scheint ein Wort keinen Inhalt zu haben, für welches uns ein entsprechendes inneres Bild fehlt. Man muss aber immer einen vollständigen Satz ins Auge fassen. Nur in ihm haben die Wörter eigentlich eine Bedeutung. Die innern Bilder, die uns dabei etwa vorschweben, brauchen nicht den logischen Bestandtheilen des Urtheils zu entsprechen. Es genügt, wenn der Satz als Ganzes einen Sinn hat; dadurch erhalten auch seine Theile ihren Inhalt.

Diese Bemerkung scheint mir geeignet, auf manche

reason for doubting the correctness of the calculation which established the distance, not does it prevent us in any way from taking that distance as a fact upon which to base further inferences.

§ 60. Even so concrete a thing as the Earth we are unable to imagine as we know it to be; instead, we content ourselves with a ball of moderate size, which serves us as a symbol for the Earth, though we know quite well it is very different from it. Thus even although our idea often fails entirely to coincide with what we want, we still make judgements about an object such as the Earth with considerable certainty, even where its size is in point.

Time and time again we are led by our thought beyond the scope of our imagination, without thereby forfeiting the support we need for our inferences. Even if, as seems to be the case, it is impossible for men such as we are to think without ideas, it is still possible for their connexion with what we are thinking of to be entirely superficial, arbitrary and conventional.

That we can form no idea of its content is therefore no reason for denying all meaning to a word, or for excluding it from our vocabulary. We are indeed only imposed on by the opposite view because we will, when asking for the meaning of a word, consider it in isolation, which leads us to accept an idea as the meaning. Accordingly, any word for which we can find no corresponding mental picture appears to have no content. But we ought always to keep before our eyes a complete proposition. Only in a proposition have the words really a meaning. It may be that mental pictures float before us all the while, but these need not correspond to the logical elements in the judgement. It is enough if the proposition taken as a whole has a sense; it is this that confers on its parts also their content.

This observation is destined, I believe, to throw light

schwierige Begriffe wie den des Unendlichkleinen*) ein Licht zu werfen, und ihre Tragweite beschränkt sich wohl nicht auf die Mathematik.

Die Selbständigkeit, die ich für die Zahl in Anspruch nehme, soll nicht bedeuten, dass ein Zahlwort ausser dem Zusammenhange eines Satzes etwas bezeichne, sondern ich will damit nur dessen Gebrauch als Praedicat oder Attribut ausschliessen, wodurch seine Bedeutung etwas verändert wird.

§ 61. Aber, wendet man vielleicht ein, mag auch die Erde eigentlich unvorstellbar sein, so ist sie doch ein äusseres Ding, das einen bestimmten Ort hat; aber wo ist die Zahl 4? sie ist weder ausser uns noch in uns. Das ist in räumlichem Sinne verstanden richtig. Eine Ortsbestimmung der Zahl 4 hat keinen Sinn; aber daraus folgt nur, dass sie kein räumlicher Gegenstand ist, nicht, dass sie überhaupt keiner ist. Nicht jeder Gegenstand ist irgendwo. Auch unsere Vorstellungen**) sind in diesem Sinne nicht in uns (subcutan). Da sind Ganglienzellen, Blutkörperchen und dergl., aber keine Vorstellungen. Räumliche Praedicate sind auf sie nicht anwendbar: die eine ist weder rechts noch links von der andern; Vorstellungen haben keine in Millimetern angebbaren Entfernungen von einander. Wenn wir sie dennoch in uns nennen, so wollen wir sie damit als subjectiv bezeichnen.

Aber wenn auch das Subjective keinen Ort hat, wie ist es möglich, dass die objective Zahl 4 nirgendwo sei? Nun ich behaupte, dass darin gar kein Widerspruch liegt. Sie ist in der That genau dieselbe für jeden, der sich mit ihr beschäftigt; aber dies hat mit Räumlichkeit nichts zu schaffen. Nicht jeder objective Gegenstand hat einen Ort.

---

*) Es kommt darauf an, den Sinn einer Gleichung wie

$$d\,f\,(x) = g\,(x)\,d\,x$$

zu definiren, nicht aber darauf, eine von zwei verschiedenen Punkten begrenzte Strecke aufzuweisen, deren Länge d x wäre.

**) Dies Wort rein psychologisch, nicht psychophysisch verstanden.

on quite a number of difficult concepts, among them that of the infinitesimal,[1] and its scope is not restricted to mathematics either.

The self-subsistence which I am claiming for number is not to be taken to mean that a number word signifies something when removed from the context of a proposition, but only to preclude the use of such words as predicates or attributes, which appreciably alters their meaning.

§ 61. But, it will perhaps be objected, even if the Earth is really not imaginable, it is at any rate an external thing, occupying a definite place; but where is the number 4? It is neither outside us nor within us. And, taking those words in their spatial sense, that is quite correct. To give spatial co-ordinates for the number 4 makes no sense; but the only conclusion to be drawn from that is, that 4 is not a spatial object, not that it is not an object at all. Not every object has a place. Even our ideas[2] are in this sense not within us (beneath our skin); beneath the skin are nerve-ganglia, blood corpuscles and things of that sort, but not ideas. Spatial predicates are not applicable to them: an idea is neither to the right nor to the left of another idea; we cannot give distances between ideas in millimetres. If we still say they are within us, then we intend by this to signify that they are subjective.

Yet even granted that what is subjective has no position in space, how is it possible for the number 4, which is objective, not to be anywhere? Now I contend that there is no contradiction in this whatever. It is a fact that the number 4 is exactly the same for everyone who deals with it; but that has nothing to do with being spatial. Not every objective object* has a place.

---

[1] The problem here is not, as might be thought, to produce a segment bounded by two distinct points whose length is dx, but rather to define the sense of an identity of the type

$$df(x) = g(x)dx$$

[2] Understanding this word in its purely psychological, not in its psycho-physical, sense.

* [*objektiver Gegenstand*]

73

Um den Begriff der Anzahl zu gewinnen, muss man
den Sinn einer Zahlengleichung feststellen.

§ 62. Wie soll uns denn eine Zahl gegeben sein, wenn
wir keine Vorstellung oder Anschauung von ihr haben können?
Nur im Zusammenhange eines Satzes bedeuten die Wörter
etwas. Es wird also darauf ankommen, den Sinn eines
Satzes zu erklären, in dem ein Zahlwort vorkommt. Das
giebt zunächst noch viel der Willkühr anheim. Aber wir
haben schon festgestellt, dass unter den Zahlwörtern selbständige Gegenstände zu verstehen sind. Damit ist uns eine
Gattung von Sätzen gegeben, die einen Sinn haben müssen,
der Sätze, welche ein Wiedererkennen ausdrücken. Wenn
uns das Zeichen a einen Gegenstand bezeichnen soll, so
müssen wir ein Kennzeichen haben, welches überall entscheidet, ob b dasselbe sei wie a, wenn es auch nicht immer
in unserer Macht steht, dies Kennzeichen anzuwenden. In
unserm Falle müssen wir den Sinn des Satzes

„die Zahl, welche dem Begriffe F zukommt, is dieselbe,
welche dem Begriffe G zukommt"

erklären; d. h. wir müssen den Inhalt dieses Satzes in anderer Weise wiedergeben, ohne den Ausdruck

„die Anzahl, welche dem Begriffe F zukommt"

zu gebrauchen. Damit geben wir ein allgemeines Kennzeichen für die Gleichheit von Zahlen an. Nachdem wir so
ein Mittel erlangt haben, eine bestimmte Zahl zu fassen
und als dieselbe wiederzuerkennen, können wir ihr ein
Zahlwort zum Eigennamen geben.

§ 63. Ein solches Mittel nennt schon Hume*): „Wenn
zwei Zahlen so combinirt werden, dass die eine immer eine
Einheit hat, die jeder Einheit der andern entspricht, so geben
wir sie als gleich an." Es scheint in neuerer Zeit die

---

*) Baumann a. a. O. Bd. II. S. 565.

To obtain the concept of Number, we must fix the sense of a numerical identity.

§ 62. How, then, are numbers to be given to us, if we cannot have any ideas or intuitions of them? Since it is only in the context of a proposition that words have any meaning, our problem becomes this: To define the sense of a proposition in which a number word occurs. That, obviously, leaves us still a very wide choice. But we have already settled that number words are to be understood as standing for self-subsistent objects. And that is enough to give us a class of propositions which must have a sense, namely those which express our recognition of a number as the same again. If we are to use the symbol *a* to signify an object, we must have a criterion for deciding in all cases whether *b* is the same as *a*, even if it is not always in our power to apply this criterion. In our present case, we have to define the sense of the proposition

"the number which belongs to the concept *F* is the same
as that which belongs to the concept *G*";

that is to say, we must reproduce the content of this proposition in other terms, avoiding the use of the expression

"the Number which belongs to the concept *F*".

In doing this, we shall be giving a general criterion for the identity of numbers. When we have thus acquired a means of arriving at a determinate number and of recognizing it again as the same, we can assign it a number word as its proper name.

§ 63. HUME[1] long ago mentioned such a means: "When two numbers are so combined as that the one has always an unit answering to every unit of the other, we pronounce them equal." This opinion, that numerical equality or identity

[1] Baumann, op. cit., Vol. II, p. 565 *Treatise*, Bk. I, Part iii, Sect. 1].

Meinung unter den Mathematikern*) vielfach Anklang gefunden zu haben, dass die Gleichheit der Zahlen mittels der eindeutigen Zuordnung definirt werden müsse. Aber es erheben sich zunächst logische Bedenken und Schwierigkeiten, an denen wir nicht ohne Prüfung vorbeigehen dürfen.

Das Verhältniss der Gleichheit kommt nicht nur bei Zahlen vor. Daraus scheint zu folgen, dass es nicht für diesen Fall besonders erklärt werden darf. Man sollte denken, dass der Begriff der Gleichheit schon vorher feststände, und dass dann aus ihm und dem Begriffe der Anzahl sich ergeben müsste, wann Anzahlen einander gleich wären, ohne dass es dazu noch einer besondern Definition bedürfte.

Hiergegen ist zu bemerken, dass für uns der Begriff der Anzahl noch nicht feststeht, sondern erst mittels unserer Erklärung bestimmt werden soll. Unsere Absicht ist, den Inhalt eines Urtheils zu bilden, der sich so als eine Gleichung auffassen lässt, dass jede Seite dieser Gleichung eine Zahl ist. Wir wollen also nicht die Gleichheit eigens für diesen Fall erklären, sondern mittels des schon bekannten Begriffes der Gleichheit, das gewinnen, was als gleich zu betrachten ist. Das scheint freilich eine sehr ungewöhnliche Art der Definition zu sein, welche wohl von den Logikern noch nicht genügend beachtet ist; dass sie aber nicht unerhört ist, mögen einige Beispiele zeigen.

§ 64. Das Urtheil: „die Gerade a ist parallel der Gerade b," in Zeichen:

$$a \; / \; / \; b,$$

kann als Gleichung aufgefasst werden. Wenn wir dies thun, erhalten wir den Begriff der Richtung und sagen: „die Richtung der Gerade a ist gleich der Richtung der Gerade b".

*) Vergl. E. Schröder a. a. O. S. 7 und 8. E. Kossak, die Elemente der Arithmetik, Programm des Friedrichs-Werder'schen Gymnasiums. Berlin, 1872. S. 16. G. Cantor, Grundlagen einer allgemeinen Mannichfaltigkeitslehre. Leipzig, 1883.

must be defined in terms of one-one correlation, seems in recent years to have gained widespread acceptance among mathematicians.[1] But it raises at once certain logical doubts and difficulties, which ought not to be passed over without examination.

It is not only among numbers that the relationship of identity is found. From which it seems to follow that we ought not to define it specially for the case of numbers. We should expect the concept of identity to have been fixed first, and that then, from it together with the concept of Number, it must be possible to deduce when Numbers are identical with one another, without there being need for this purpose of a special definition of numerical identity as well.

As against this, it must be noted that for us the concept of Number has not yet been fixed, but is only due to be determined in the light of our definition of numerical identity. Our aim is to construct the content of a judgement which can be taken as an identity such that each side of it is a number. We are therefore proposing not to define identity specially for this case, but to use the concept of identity, taken as already known, as a means for arriving at that which is to be regarded as being identical. Admittedly, this seems to be a very odd kind of definition, to which logicians have not yet paid enough attention; but that it is not altogether unheard of, may be shown by a few examples.

§ 64. The judgement "line a is parallel to line b", or, using symbols,

$$a \mathbin{/\!/} b,$$

can be taken as an identity. If we do this, we obtain the concept of direction, and say: "the direction of line a is identical with the direction of line b". Thus we replace the symbol $/\!/$ by

---

[1] Cf. E. Schröder, op. cit., pp. 7-8; E. Kossak, *Die Elemente der Arithmetik, Programm des Friedrichs-Werder'schen Gymnasiums*, Berlin 1872, p. 16; G. Cantor, *Grundlagen einer allgemeinen Mannichfaltigkeitslehre*, Leipzig 1883.

Wir ersetzen also das Zeichen / / durch das allgemeinere $=$' indem wir den besondern Inhalt des ersteren an a und b vertheilen. Wir zerspalten den Inhalt in anderer als der ursprünglichen Weise und gewinnen dadurch einen neuen Begriff. Oft fasst man freilich die Sache umgekehrt auf, und manche Lehrer definiren: parallele Geraden sind solche von gleicher Richtung. Der Satz: „wenn zwei Geraden einer dritten parallel sind, so sind sie einander parallel" lässt sich dann mit Berufung auf den ähnlich lautenden Gleichheitssatz sehr bequem beweisen. Nur schade, dass der wahre Sachverhalt damit auf den Kopf gestellt wird! Denn alles Geometrische muss doch wohl ursprünglich anschaulich sein. Nun frage ich, ob jemand eine Anschauung von der Richtung einer Gerade hat. Von der Gerade wohl! aber unterscheidet man in der Anschauung von dieser Gerade noch ihre Richtung? Schwerlich! Dieser Begriff wird erst durch eine an die Anschauung anknüpfende geistige Thätigkeit gefunden. Dagegen hat man eine Vorstellung von parallelen Geraden. Jener Beweis kommt nur durch eine Erschleichung zu Stande, indem man durch den Gebrauch des Wortes „Richtung" das zu Beweisende voraussetzt; denn wäre der Satz: „wenn zwei Geraden einer dritten parallel sind, so sind sie einander parallel" unrichtig, so könnte man a//b nicht in eine Gleichung verwandeln.

So kann man aus dem Parallelismus von Ebenen einen Begriff erhalten, der dem der Richtung bei Geraden entspricht. Ich habe dafür den Namen „Stellung" gelesen. Aus der geometrischen Aehnlichkeit geht der Begriff der Gestalt hervor, so dass man z. B. statt „die beiden Dreiecke sind ähnlich" sagt: „die beiden Dreiecke haben gleiche Gestalt" oder „die Gestalt des einen Dreiecks ist gleich der Gestalt des andern". So kann man auch aus der collinearen Verwandtschaft geometrischer Gebilde einen Begriff gewinnen, für den ein Name wohl noch fehlt.

the more generic symbol =, through removing what is specific in the content of the former and dividing it between *a* and *b*. We carve up the content in a way different from the original way, and this yields us a new concept. Often, of course, we conceive of the matter the other way round, and many authorities define parallel lines as lines whose directions are identical. The proposition that "straight lines parallel to the same straight line are parallel to one another" can than be very conveniently proved by invoking the analogous proposition about things identical with the same thing. Only the trouble is, that this is to reverse the true order of things. For surely everything geometrical must be given originally in intuition. But now I ask whether anyone has an intuition of the direction of a straight line. Of a straight line, certainly; but do we distinguish in our intuition between this straight line and something else, its direction? That is hardly plausible. The concept of direction is only discovered at all as a result of a process of intellectual activity which takes its start from the intuition. On the other hand, we do have an idea of parallel straight lines. Our convenient proof is only made possible by surreptitiously assuming, in our use of the word "direction", what was to be proved; for if it were false that "straight lines parallel to the same straight line are parallel to one another", then we could not transform *a* / / *b* into an identity.

We can obtain in a similar way from the parallelism of planes a concept corresponding to that of direction in the case of straight lines; I have seen the name "orientation"* used for this. From geometrical similarity is derived the concept of shape, so that instead of "the two triangles are similar" we say "the two triangles are of identical shape" or "the shape of the one is identical with that of the other". It is possible to derive yet another concept in this way, to which no name has yet been given, from the collineation of geometrical forms.

* [*Stellung*]

§ 65. Um nun z. B. vom Parallelismus*) auf den Begriff der Richtung zu kommen, versuchen wir folgende Definition: der Satz

„die Gerade a ist parallel der Gerade b"

sei gleichbedeutend mit

„die Richtung der Gerade a ist gleich der Richtung
der Gerade b".

Diese Erklärung weicht insofern von dem Gewohnten ab, als sie scheinbar die schon bekannte Beziehung der Gleichheit bestimmt, während sie in Wahrheit den Ausdruck „die Richtung der Gerade a" einführen soll, der nur nebensächlich vorkommt. Daraus entspringt ein zweites Bedenken, ob wir nicht durch eine solche Festsetzung in Widersprüche mit den bekannten Gesetzen der Gleichheit verwickelt werden könnten. Welches sind diese? Sie werden als analytische Wahrheiten aus dem Begriffe selbst entwickelt werden können. Nun definirt Leibniz**):

„Eadem sunt, quorum unum potest substitui alteri
salva veritate."

Diese Erklärung eigne ich mir für die Gleichheit an. Ob man wie Leibniz „dasselbe" sagt oder „gleich", ist unerheblich. „Dasselbe" scheint zwar eine vollkommene Uebereinstimmung „gleich" nur eine in dieser oder jener Hinsicht auszudrücken; man kann aber eine solche Redeweise annehmen, dass dieser Unterschied wegfällt, indem man z. B. statt „die Strecken sind in der Länge gleich" sagt „die Länge der Strecken ist gleich" oder „dieselbe," statt „die Flächen sind in der Farbe gleich" „die Farbe der Flächen ist gleich". Und so haben wir das Wort oben in den Beispielen

---

*) Um mich bequemer ausdrücken zu können und leichter verstanden zu werden, spreche ich hier vom Parallelismus. Das Wesentliche dieser Erörterungen wird leicht auf den Fall der Zahlengleichheit übertragen werden können.

**) Non inelegans specimen demonstrandi in abstractis. Erdm. S. 94.

§ 65. Now in order to get, for example, from parallelism[1]
to the concept of direction, let us try the following definition:
The proposition

"line a is parallel to line b"

is to mean the same as

"the direction of line a is identical with the direction of
line b".

This definition departs to some extent from normal
practice, in that it serves ostensibly to adapt the relation of
identity, taken as already known, to a special case, whereas
in reality it is designed to introduce the expression "the
direction of line a", which only comes into it incidentally. It is
this that gives rise to a second doubt—are we not liable,
through using such methods, to become involved in conflict
with the well-known laws of identity? Let us see what these
are. As analytic truths they should be capable of being derived
from the concept itself alone. Now LEIBNIZ's[2] definition is as
follows:

"Things are the same as each other, of which one can
be substituted for the other without loss of truth".*

This I propose to adopt as my own definition of identity.
Whether we use "the same", as LEIBNIZ does, or "identical", is
not of any importance. "The same" may indeed be thought
to refer to complete agreement in all respects, "identical"**
only to agreement in this respect or that; but we can adopt a
form of expression such that this distinction vanishes. For
example, instead of "the segments are identical in length",
we can say "the length of the segments is identical" or "the
same", and instead of "the surfaces are identical in colour",
"the colour of the surfaces is identical". And this is the way
in which the word has been used in the examples above.

---

[1] I have chosen to discuss here the case of parallelism, because I can express
myself less clumsily and make myself more easily understood. The argument
can readily be transferred in essentials to apply to the case of numerical identity.

[2] *Non inelegans specimen demonstrandi in abstractis* (Erdmann edn., p. 94).

* [*Eadem sunt, quorum unum potest substitui alteri salva veritate.*]

** [Still more "equal" or "similar", which the German *gleich* can also mean.]

gebraucht. In der allgemeinen Ersetzbarkeit sind nun in der That alle Gesetze der Gleichheit enthalten.

Um unsern Definitionsversuch der Richtung einer Gerade zu rechtfertigen, müssten wir also zeigen, dass man

die Richtung von a

überall durch

die Richtung von b

ersetzen könne, wenn die Gerade a der Gerade b parallel ist. Dies wird dadurch vereinfacht, dass man zunächst von der Richtung einer Gerade keine andere Aussage kennt als die Uebereinstimmung mit der Richtung einer andern Gerade. Wir brauchten also nur die Ersetzbarkeit in einer solchen Gleichheit nachzuweisen oder in Inhalten, welche solche Gleichheiten als Bestandtheile*) enthalten würden. Alle andern Aussagen von Richtungen müssten erst erklärt werden und für diese Definitionen können wir die Regel aufstellen, dass die Ersetzbarkeit der Richtung einer Gerade durch die einer ihr parallelen gewahrt bleiben muss.

§ 66. Aber noch ein drittes Bedenken erhebt sich gegen unsern Definitionsversuch. In dem Satze

„die Richtung von a ist gleich der Richtung von b"

erscheint die Richtung von a als Gegenstand**) und wir haben in unserer Definition ein Mittel, diesen Gegenstand wiederzuerkennen, wenn er etwa in einer andern Verkleidung etwa als Richtung von b auftreten sollte. Aber dies Mittel

---

*) In einem hypothetischen Urtheile könnte z. B. eine Gleichheit von Richtungen als Bedingung oder Folge vorkommen.

**) Der bestimmte Artikel deutet dies an. Begriff ist für mich ein mögliches Praedicat eines singulären beurtheilbaren Inhalts, Gegenstand ein mögliches Subject eines solchen. Wenn wir in dem Satze

„die Richtung der Fernrohraxe ist gleich der Richtung der Erdaxe"

die Richtung der Fernrohraxe als Subject ansehen, so ist das Praedicat „gleich der Richtung der Erdaxe". Dies ist ein Begriff. Aber die Richtung der Erdaxe ist nur ein Theil des Praedicates; sie ist ein Gegenstand, da sie auch zum Subjecte gemacht werden kann.

Now, it is actually the case that in universal substitutability all the laws of identity are contained.

In order, therefore, to justify our proposed definition of the direction of a line, we should have to show that it is possible, if line *a* is parallel to line *b*, to substitute

"the direction of *b*"

everywhere for

"the direction of *a*".

This task is made simpler by the fact that we are being taken initially to know of nothing that can be asserted about the direction of a line except the one thing, that it coincides with the direction of some other line. We should thus have to show only that substitution was possible in an identity of this one type, or in judgement-contents containing such identities as constituent elements.[1] The meaning of any other type of assertion about directions would have first of all to be defined, and in defining it we can make it a rule always to see that it must remain possible to substitute for the direction of any line the direction of any line parallel to it.

§ 66. But there is still a third doubt which may make us suspicious of our proposed definition. In the proposition

"the direction of *a* is identical with the direction of *b*"

the direction of *a* plays the part of an object,[2] and our definition affords us a means of recognizing this object as the same again, in case it should happen to crop up in some other guise, say as the direction of *b*. But this means does not provide for all

---

[1] In a hypothetical judgement, for example, an identity of directions might occur as antecedent or consequent.

[2] This is shown by the definite article. A concept is for me that which can be predicate of a singular judgement-content, an object that which can be subject of the same. If in the proposition

"the direction of the axis of the telescope is identical with the direction
of the Earth's axis"

we take the direction of the axis of the telescope as subject, then the predicate is "identical with the direction of the Earth's axis". This is a concept. But the direction of the Earth's axis is only an element in the predicate; it, since it can also be made the subject, is an object.

reicht nicht für alle Fälle aus. Man kann z. B. danach nicht entscheiden, ob England dasselbe sei wie die Richtung der Erdaxe. Man verzeihe dies unsinnig scheinende Beispiel! Natürlich wird niemand England mit der Richtung der Erdaxe verwechseln; aber dies ist nicht das Verdienst unserer Erklärung. Diese sagt nichts darüber, ob der Satz

„die Richtung von a ist gleich q"

zu bejahen oder zu verneinen ist, wenn nicht q selbst in der Form „die Richtung von b" gegeben ist. Es fehlt uns der Begriff der Richtung; denn hätten wir diesen, so könnten wir festsetzen; wenn q keine Richtung ist, so ist unser Satz zu verneinen; wenn q eine Richtung ist, so entscheidet die frühere Erklärung. Es liegt nun nahe zu erklären:

q ist eine Richtung, wenn es eine Gerade b giebt,

deren Richtung q ist.

Aber nun ist klar, dass wir uns im Kreise gedreht haben. Um diese Erklärung anwenden zu können, müssen wir schon in jedem Falle wissen, ob der Satz

„q ist gleich der Richtung von b"

zu bejahen oder zu verneinen wäre.

§ 67. Wenn man sagen wollte: q ist eine Richtung, wenn es durch die oben ausgesprochene Definition eingeführt ist, so würde man die Weise, wie der Gegenstand q eingeführt ist, als dessen Eigenschaft behandeln, was sie nicht ist. Die Definition eines Gegenstandes sagt als solche eigentlich nichts von ihm aus, sondern setzt die Bedeutung eines Zeichens fest. Nachdem das geschehen ist, verwandelt sie sich in ein Urtheil, das von dem Gegenstande handelt, aber führt ihn nun auch nicht mehr ein und steht mit andern Aussagen von ihm in gleicher Linie. Man würde, wenn man diesen Ausweg wählte, voraussetzen, dass ein Gegenstand nur auf eine einzige Weise gegeben werden könnte; denn sonst würde daraus, dass q nicht durch unsere Definition eingeführt ist, nicht folgen, dass es nicht so eingeführt werden könnte. Alle Gleichungen würden darauf hinaus-

cases. It will not, for instance, decide for us whether England is the same as the direction of the Earth's axis—if I may be forgiven an example which looks nonsensical. Naturally no one is going to confuse England with the direction of the Earth's axis; but that is no thanks to our definition of direction. That says nothing as to whether the proposition

"the direction of a is identical with q"

should be affirmed or denied, except for the one case where q is given in the form of "the direction of b". What we lack is the concept of direction; for if we had that, then we could lay it down that, if q is not a direction, our proposition is to be denied, while if it is a direction, our original definition will decide whether it is to be denied or affirmed. So the temptation is to give as our definition:

q is a direction, if there is a line b whose direction is q.

But then we have obviously come round in a circle. For in order to make use of this definition, we should have to know already in every case whether the proposition

"q is identical with the direction of b"

was to be affirmed or denied.

§ 67. If we were to try saying: q is a direction if it is introduced by means of the definition set out above, then we should be treating the way in which the object q is introduced as a property of q, which it is not. The definition of an object does not, as such, really assert anything about the object, but only lays down the meaning of a symbol. After this has been done, the definition transforms itself into a judgement, which does assert about the object; but now it no longer introduces the object, it is exactly on a level with other assertions made about it. If, moreover, we were to adopt this way out, we should have to be presupposing that an object can only be given in one single way; for otherwise it would not follow, from the fact that q *was* not introduced by means of our definition, that it *could* not have been introduced by means of it. All identities would then amount simply to this, that whatever

kommen, dass das als dasselbe anerkannt würde, was uns auf dieselbe Weise gegeben ist. Aber dies ist so selbstverständlich und so unfruchtbar, dass es nicht verlohnte, es auszusprechen. Man könnte in der That keinen Schluss daraus ziehen, der von jeder der Voraussetzungen verschieden wäre. Die vielseitige und bedeutsame Verwendbarkeit der Gleichungen beruht vielmehr darauf, dass man etwas wiedererkennen kann, obwohl es auf verschiedene Weise gegeben ist.

§ 68. Da wir so keinen scharf begrenzten Begriff der Richtung und aus denselben Gründen keinen solchen der Anzahl gewinnen können, versuchen wir einen andern Weg. Wenn die Gerade a der Gerade b parallel ist, so ist der Umfang des Begriffes „Gerade parallel der Gerade a" gleich dem Umfange des Begriffes „Gerade parallel der Gerade b"; und umgekehrt: wenn die Umfänge der genannten Begriffe gleich sind, so ist a parallel b. Versuchen wir also zu erklären:

die Richtung der Gerade a ist der Umfang des Begriffes „parallel der Gerade a";
die Gestalt des Dreiecks d ist der Umfang des Begriffes „ähnlich dem Dreiecke d"!

Wenn wir dies auf unsern Fall anwenden wollen, so haben wir an die Stelle der Geraden oder der Dreiecke Begriffe zu setzen und an die Stelle des Parallelismus oder der Aehnlichkeit die Möglichkeit die unter den einen den unter den andern Begriff fallenden Gegenständen beiderseits eindeutig zuzuordnen. Ich will der Kürze wegen den Begriff F dem Begriffe G gleichzahlig nennen, wenn diese Möglichkeit vorliegt, muss aber bitten, dies Wort als eine willkührlich gewählte Bezeichnungsweise zu betrachten, deren Bedeutung nicht der sprachlichen Zusammensetzung, sondern dieser Festsetzung zu entnehmen ist.

Ich definire demnach:

die Anzahl, welche dem Begriffe F zukommt, ist

is given to us in the same way is to be reckoned as the same. This, however, is a principle so obvious and so sterile as not to be worth stating. We could not, in fact, draw from it any conclusion which was not the same as one of our premisses. Why is it, after all, that we are able to make use of identities with such significant results in such divers fields? Surely it is rather because we are able to recognize something as the same again even although it is given in a different way.

§ 68. Seeing that we cannot by these methods obtain any concept of direction with sharp limits to its application, nor therefore, for the same reasons, any satisfactory concept of Number either, let us try another way. If line *a* is parallel to line *b*, then the extension of the concept "line parallel to line *a*" is identical with the extension of the concept "line parallel to line *b*"; and conversely, if the extensions of the two concepts just named are identical, then *a* is parallel to *b*. Let us try, therefore, the following type of definition:

> the direction of line *a* is the extension of the concept "parallel to line *a*";
> the shape of triangle *t* is the extension of the concept "similar to triangle *t*".

To apply this to our own case of Number, we must substitute for lines or triangles concepts, and for parallelism or similarity the possibility of correlating one to one the objects which fall under the one concept with those which fall under the other. For brevity, I shall, when this condition is satisfied, speak of the concept *F* being *equal** to the concept *G*; but I must ask that this word be treated as an arbitrarily selected symbol, whose meaning is to be gathered, not from its etymology, but from what is here laid down.

My definition is therefore as follows:

> the Number which belongs to the concept *F* is the

---

* [*Gleichzahlig*—an invented word, literally "identinumerate" or "taut-arithmic"; but these are too clumsy for constant use. Other translators have used "equinumerous"; "equinumerate" would be better. Later writers have used "similar" in this connexion (but as a predicate of "class" not of "concept").]

der Umfang*) des Begriffes „gleichzahlig dem Begriffe F"

§ 69. Dass diese Erklärung zutreffe, wird zunächst
vielleicht wenig einleuchten. Denkt man sich unter dem
Umfange eines Begriffes nicht etwas Anderes? Was man
sich darunter denkt, erhellt aus den ursprünglichen Aussagen,
die von Begriffsumfängen gemacht werden können. Es sind
folgende:

1. die Gleichheit,
2. dass der eine umfassender als der andere sei.

Nun ist der Satz:

> der Umfang des Begriffes „gleichzahlig dem Begriffe
> F" ist gleich dem Umfange des Begriffes „gleichzahlig
> dem Begriffe G"

immer dann und nur dann wahr, wenn auch der Satz

> „dem Begriffe F kommt dieselbe Zahl wie dem Begriffe
> G zu"

wahr ist. Hier ist also voller Einklang.

Man sagt zwar nicht, dass eine Zahl umfassender als
eine andere sei in dem Sinne, wie der Umfang eines Begriffes
umfassender als der eines andern ist; aber der Fall, dass

> der Umfang des Begriffes „gleichzahlig dem Be-
> griffe F"

umfassender sei als

---

*) Ich glaube, dass für „Umfang des Begriffes" einfach „Begriff" gesagt
werden könnte. Aber man würde zweierlei einwenden:

1. dies stehe im Widerspruche mit meiner früheren Behauptung dass
die einzelne Zahl ein Gegenstand sei, was durch den bestimmten Artikel
in Ausdrücken wie „die Zwei" und durch die Unmöglichkeit angedeutet
werde, von Einsen, Zweien u. s. w. im Plural zu sprechen, sowie
dadurch, dass die Zahl nur einen Theil des Praedicats der Zahlangabe aus-
mache;

2. dass Begriffe von gleichem Umfange sein können, ohne zu-
sammenzufallen.

Ich bin nun zwar der Meinung, dass beide Einwände gehoben
werden können; aber das möchte hier zu weit führen. Ich setze voraus,
dass man wisse, was der Umfang eines Begriffes sei.

extension[1] of the concept "equal to the concept *F*".

§ 69. That this definition is correct will perhaps be hardly evident at first. For do we not think of the extensions of concepts as something quite different from numbers? How we do think of them emerges clearly from the basic assertions we make about them. These are as follows:

    1. that they are identical,
    2. that one is wider than the other.

But now the proposition:

the extension of the concept "equal to the concept *F*" is identical with the extension of the concept "equal to the concept *G*"

is true if and only if the proposition

"the same number belongs to the concept *F* as to the concept *G*"

is also true. So that here there is complete agreement.

Certainly we do not say that one number is wider than another, in the sense in which the extension of one concept is wider than that of another; but then it is also quite impossible for a case to occur where

the extension of the concept "equal to the concept *F*"

would be wider than

---

[1] I believe that for "extension of the concept" we could write simply "concept". But this would be open to the two objections:

    1. that this contradicts my earlier statement that the individual numbers are objects, as is indicated by the use of the definite article in expressions like "the number two" and by the impossibility of speaking of ones, twos, etc. in the plural, as also by the fact that the number constitutes only an element in the predicate of a statement of number;

    2. that concepts can have identical extensions without themselves coinciding.

I am, as it happens, convinced that both these objections can be met; but to do this would take us too far afield for present purposes. I assume that it is known what the extension of a concept is.

der Umfang des Begriffes „gleichzahlig dem Begriffe G" kann auch gar nicht vorkommen; sondern, wenn alle Begriffe, die dem G gleichzahlig sind, auch dem F gleichzahlig sind, so sind auch umgekehrt alle Begriffe, die dem F gleichzahlig sind, dem G gleichzahlig. Dies „umfassender" darf natürlich nicht mit dem „grösser" verwechselt werden, dass bei Zahlen vorkommt.

Freilich ist noch der Fall denkbar, dass der Umfang des Begriffes „gleichzahlig dem Begriffe F" umfassender oder weniger umfassend wäre als ein anderer Begriffsumfang, der dann nach unserer Erklärung keine Anzahl sein könnte; und es ist nicht üblich, eine Anzahl umfassender oder weniger umfassend als den Umfang eines Begriffes zu nennen; aber es steht auch nichts im Wege, eine solche Redeweise anzunehmen, falls solches einmal vorkommen sollte.

Ergänzung und Bewährung unserer Definition.

§ 70. Definitionen bewähren sich durch ihre Fruchtbarkeit. Solche, die ebensogut wegbleiben könnten, ohne eine Lücke in der Beweisführung zu öffnen, sind als völlig werthlos zu verwerfen.

Versuchen wir also, ob sich bekannte Eigenschaften der Zahlen aus unserer Erklärung der Anzahl, welche dem Begriffe F zukommt, ableiten lassen! Wir werden uns hier mit den einfachsten begnügen.

Dazu ist es nöthig, die Gleichzahligkeit noch etwas genauer zu fassen. Wir erklärten sie mittels der beiderseits eindeutigen Zuordnung, und wie ich diesen Ausdruck verstehen will, ist jetzt darzulegen, weil man leicht etwas Anschauliches darin vermuthen könnte.

Betrachten wir folgendes Beispiel! Wenn ein Kellner sicher sein will, dass er ebensoviele Messer als Teller auf den Tisch legt, braucht er weder diese noch jene zu zählen,

the extension of the concept "equal to the concept G". For on the contrary, when all concepts equal to G are also equal to F, then conversely also all concepts equal to F are equal to G. "Wider" as used here must not, of course, be confused with "greater" as used of numbers.

Another type of case is, I admit, conceivable, where the extension of the concept "equal to the concept F" might be wider or less wide than the extension of some other concept, which then could not, on our definition, be a Number; and it is not usual to speak of a Number as wider or less wide than the extension of a concept; but neither is there anything to prevent us speaking in this way, if such a case should ever occur.

Our definition completed and its worth proved.

§ 70. Definitions show their worth by proving fruitful. Those that could just as well be omitted and leave no link missing in the chain of our proofs should be rejected as completely worthless.

Let us try, therefore, whether we can derive from our definition of the Number which belongs to the concept F any of the well-known properties of numbers. We shall confine ourselves here to the simplest.

For this it is necessary to give a rather more precise account still of the term "equality". "Equal" we defined in terms of one-one correlation, and what must now be laid down is how this latter expression is to be understood, since it might easily be supposed that it had something to do with intuition.

We will consider the following example. If a waiter wishes to be certain of laying exactly as many knives on a table as plates, he has no need to count either of them; all he

wenn er nur rechts neben jeden Teller ein Messer legt, sodass jedes Messer auf dem Tische sich rechts neben einem
Teller befindet. Die Teller und Messer sind so beiderseits
eindeutig einander zugeordnet und zwar durch das gleiche
Lagenverhältniss. Wenn wir in dem Satze

„a liegt rechts neben A"

für a und A andere und andere Gegenstände eingesetzt
denken, so macht der hierbei unverändert bleibende Theil
des Inhalts das Wesen der Beziehung aus. Verallgemeinern
wir dies!

Indem wir von einem beurtheilbaren Inhalte, der von
einem Gegenstande a und von einem Gegenstande b handelt,
a und b absondern, so behalten wir einen Beziehungsbegriff
übrig, der demnach in doppelter Weise ergänzungsbedürftig
ist. Wenn wir in dem Satze:

„die Erde hat mehr Masse als der Mond"

„die Erde" absondern, so erhalten wir den Begriff „mehr
Masse als der Mond habend". Wenn wir dagegen den
Gegenstand „der Mond" absondern, gewinnen wir den Begriff
„weniger Masse als die Erde habend". Sondern wir beide
zugleich ab, so bleibt ein Beziehungsbegriff zurück, der für
sich allein ebensowenig wie ein einfacher Begriff einen Sinn
hat: er verlangt immer eine Ergänzung zu einem beurtheilbaren Inhalte. Aber diese kann in verschiedener Weise
geschehen: statt Erde und Mond kann ich z. B. Sonne und
Erde setzen, und hierdurch wird eben die Absonderung
bewirkt.

Die einzelnen Paare zugeordneter Gegenstände verhalten sich in ähnlicher Weise — man könnte sagen als
Subjecte — zu dem Beziehungsbegriffe, wie der einzelne
Gegenstand zu dem Begriffe unter den er fällt. Das Subject ist hier ein zusammengesetztes. Zuweilen, wenn die
Beziehung eine umkehrbare ist, kommt dies auch sprachlich
zum Ausdrucke wie in dem Satze „Peleus und Thetis waren

has to do is to lay immediately to the right of every plate a knife, taking care that every knife on the table lies immediately to the right of a plate. Plates and knives are thus correlated one to one, and that by the identical spatial relationship. Now if in the proposition

"a lies immediately to the right of A"

we conceive first one and then another object inserted in place of a and again of A, then that part of the content which remains unaltered throughout this process constitutes the essence of the relation. What we need is a generalization of this.

If from a judgement-content which deals with an object a and an object b we subtract a and b, we obtain as remainder a relation-concept which is, accordingly, incomplete at two points. If from the proposition

"the Earth is more massive than the Moon"

we subtract "the Earth", we obtain the concept "more massive than the Moon". If, alternatively, we subtract the object, "the Moon", we get the concept "less massive than the Earth". But if we subtract them both at once, then we are left with a relation-concept, which taken by itself has no [assertible] sense any more than a simple concept has: it has always to be completed in order to make up a judgement-content. It can however be completed in different ways: instead of Earth and Moon I can put, for example, Sun and Earth, and this *eo ipso* effects the subtraction.

Each individual pair of correlated objects stands to the relation-concept much as an individual object stands to the concept under which it falls—we might call them the subject of the relation-concept. Only here the subject is a composite one. Occasionally, where the relation in question is convertible, this fact achieves verbal recognition, as in the proposi-

die Eltern des Achilleus"*). Dagegen wäre es z. B. nicht gut möglich, den Inhalt des Satzes „die Erde ist grösser als der Mond" so wiederzugeben, dass „die Erde und der Mond" als zusammengesetztes Subject erschiene, weil das „und" immer eine gewisse Gleichstellung andeutet. Aber dies thut nichts zur Sache.

Der Beziehungsbegriff gehört also wie der einfache der reinen Logik an. Es kommt hier nicht der besondere Inhalt der Beziehung in Betracht, sondern allein die logische Form. Und was von dieser ausgesagt werden kann, dessen Wahrheit ist analytisch und wird a priori erkannt. Dies gilt von den Beziehungsbegriffen wie von den andern.

Wie

„a fällt unter den Begriff F"

die allgemeine Form eines beurtheilbaren Inhalts ist, der von einem Gegenstande a handelt, so kann man

„a steht in der Beziehung ϕ zu b"

als allgemeine Form für einen beurtheilbaren Inhalt annehmen, der von dem Gegenstande a und von dem Gegenstande b handelt.

§ 71. Wenn nun jeder Gegenstand, der unter den Begriff F fällt, in der Beziehung ϕ zu einem unter den Begriff G fallenden Gegenstande steht, und wenn zu jedem Gegenstande, der unter G fällt, ein unter F fallender Gegenstand in der Beziehung ϕ steht, so sind die unter F und G fallenden Gegenstände durch die Beziehung ϕ einander zugeordnet.

Es kann noch gefragt werden, was der Ausdruck

„jeder Gegenstand, der unter F fällt, steht in der Beziehung ϕ zu einem unter G fallenden Gegenstande"

bedeute, wenn gar kein Gegenstand unter F fällt. Ich verstehe darunter:

---

*) Hiermit ist der Fall nicht zu verwechseln, wo das „und" nur scheinbar die Subjecte, in Wahrheit aber zwei Sätze verbindet.

tion "Peleus and Thetis were the parents of Achilles".[1] But not always. For example, it would scarcely be possible to put the proposition "the Earth is bigger than the Moon" into other words so as to make "the Earth and the Moon" appear as a composite subject; the "and" must always indicate that the two things are being put in some way on a level. However, this does not affect the issue.

The doctrine of relation-concepts is thus, like that of simple concepts, a part of pure logic. What is of concern to logic is not the special content of any particular relation, but only the logical form. And whatever can be asserted of this, is true analytically and known a priori. This is as true of relation-concepts as of other concepts.

Just as
    "a falls under the concept F"
is the general form of a judgement-content which deals with an object a, so we can take

    "a stands in the relation ϕ to b"

as the general form of a judgement-content which deals with an object a and an object b.

§ 71. If now every object which falls under the concept F stands in the relation ϕ to an object falling under the concept G, and if to every object which falls under G there stands in the relation ϕ an object falling under F, then the objects falling under F and under G are correlated with each other by the relation ϕ.

It may still be asked, what is the meaning of the expression

    "every object which falls under F stands in the relation ϕ to an object falling under G"

in the case where no object at all falls under F. I understand this expression as follows:

---

[1] This type of case should not be confused with another, in which the "and" joins the subjects in appearance only, but in reality joins two propositions.

die beiden Sätze

„a fällt unter F"

und

„a steht zu keinem unter G fallenden Gegenstande in der Beziehung ϕ"

können nicht mit einander bestehen, was auch a bezeichne, sodass entweder der erste oder der zweite oder beide falsch sind. Hieraus geht hervor, dass „jeder Gegenstand, der unter F fällt, in der Beziehung ϕ zu einem unter G fallenden Gegenstande steht," wenn es keinen unter F fallenden Gegenstand giebt, weil dann der erste Satz

„a fällt unter F"

immer zu verneinen ist, was auch a sein mag.

Ebenso bedeutet

„zu jedem Gegenstande, der unter G fällt, steht ein unter F fallender in der Beziehung ϕ",

dass die beiden Sätze

„a fällt unter G"

und

„kein unter F fallender Gegenstand steht zu a in der Beziehung ϕ"

nicht mit einander bestehen können, was auch a sein möge.

§ 72. Wir haben nun gesehen, wann die unter die Begriffe F und G fallenden Gegenstände einander durch die Beziehung ϕ zugeordnet sind. Hier soll nun diese Zuordnung eine beiderseits eindeutige sein. Darunter verstehe ich, dass folgende beiden Sätze gelten:

1. wenn d in der Beziehung ϕ zu a steht, und wenn d in der Beziehung ϕ zu e steht, so ist allgemein, was auch d, a und e sein mögen, a dasselbe wie e;

2. wenn d in der Beziehung ϕ zu a steht, und wenn b in der Beziehung ϕ zu a steht, so ist allgemein, was auch d, b und a sein mögen, d dasselbe wie b.

Hiermit haben wir die beiderseits eindeutige Zuordnung

the two propositions

"*a* falls under *F*"

and

"*a* does not stand in the relation ϕ to any object
falling under *G*"

cannot, whatever be signified by *a*, both be true together;
so that either the first proposition is false, or the second is, or
both are. From this it can be seen that the proposition "every
object which falls under *F* stands in the relation ϕ to an
object falling under *G*" is, in the case where there is no object
falling under *F*, true; for in that case the first proposition

"*a* falls under *F*"

is always false, whatever *a* may be.

In the same way the proposition

"to every object which falls under *G* there stands in
the relation ϕ an object falling under *F*"

means that the two propositions

"*a* falls under *G*"

and

"no object falling under *F* stands to *a* in the
relation ϕ"

cannot, whatever *a* may be, both be true together.

§ 72. We have thus seen when the objects falling under
the concepts *F* and *G* are correlated with each other by the
relation ϕ. But now in our case, this correlation has to be
one-one. By this I understand that the two following
propositions both hold good:

1. If *d* stands in the relation ϕ to *a*, and if *d* stands in the
   relation ϕ to *e*, then generally, whatever *d*, *a* and *e* may
   be, *a* is the same as *e*.
2. If *d* stands in the relation ϕ to *a*, and if *b* stands in the
   relation ϕ to *a*, then generally, whatever *d*, *b* and *a* may be,
   d is the same as *b*.

This reduces one-one correlation to purely logical

auf rein logische Verhältnisse zurückgeführt und können nun so definiren:

der Ausdruck

„der Begriff F ist gleichzahlig dem Begriffe G"

sei gleichbedeutend mit dem Ausdrucke

„es giebt eine Beziehung ϕ, welche die unter den Begriff F fallenden Gegenstände den unter G fallenden Gegenständen beiderseits eindeutig zuordnet".

Ich wiederhole:

die Anzahl, welche dem Begriffe F zukommt, ist der Umfang des Begriffes „gleichzahlig dem Begriffe F"

und füge hinzu:

der Ausdruck

„n ist eine Anzahl"

sei gleichbedeutend mit dem Ausdrucke

„es giebt einen Begriff der Art, dass n die Anzahl ist, welche ihm zukommt".

So ist der Begriff der Anzahl erklärt, scheinbar freilich durch sich selbst, aber dennoch ohne Fehler, weil „die Anzahl, welche dem Begriffe F zukommt" schon erklärt ist.

§ 73. Wir wollen nun zunächst zeigen, dass die Anzahl, welche dem Begriffe F zukommt, gleich der Anzahl ist, welche dem Begriffe G zukommt, wenn der Begriff F dem Begriffe G gleichzahlig ist. Dies klingt freilich wie eine Tautologie, ist es aber nicht, da die Bedeutung des Wortes „gleichzahlig" nicht aus der Zusammensetzung, sondern aus der eben gegebenen Erklärung hervorgeht.

Nach unserer Definition ist zu zeigen, dass der Umfang des Begriffes „gleichzahlig dem Begriffe F" derselbe ist wie der Umfang des Begriffes „gleichzahlig dem Begriffe G," wenn der Begriff F gleichzahlig dem Begriffe G ist. Mit andern Worten: es muss bewiesen werden, dass unter dieser Voraussetzung die Sätze allgemein gelten:

wenn der Begriff H gleichzahlig dem Begriffe F ist,

relationships, and enables us to give the following definition:

the expression

"the concept *F* is equal to the concept *G*"

is to mean the same as the expression

"there exists a relation ϕ which correlates one to one the objects falling under the concept *F* with the objects falling under the concept *G*".

We now repeat our original definition:

the Number which belongs to the concept *F* is the extension of the concept "equal to the concept *F*"

and add further:

the expression

"*n* is a Number"

is to mean the same as the expression

"there exists a concept such that *n* is the Number which belongs to it".

Thus the concept of Number receives its definition, apparently, indeed, in terms of itself, but actually without any fallacy, since "the Number which belongs to the concept *F*" has already been defined.

§ 73. Our next aim must be to show that the Number which belongs to the concept *F* is identical with the Number which belongs to the concept *G* if the concept *F* is equal to the concept *G*. This sounds, of course, like a tautology. But it is not; the meaning of the word "equal" is not to be inferred from its etymology, but taken to be as I defined it above.

On our definition [of "the Number which belongs to the concept *F*"], what has to be shown is that the extension of the concept "equal to the concept *F*" is the same as the extension of the concept "equal to the concept *G*", if the concept *F* is equal to the concept *G*. In other words: it is to be proved that, for *F* equal to *G*, the following two propositions hold good universally:

if the concept *H* is equal to the concept *F*,

so ist er auch gleichzahlig dem Begriffe G;

und

wenn der Begriff H dem Begriffe G gleichzahlig ist, so ist er auch gleichzahlig dem Begriffe F.

Der erste Satz kommt darauf hinaus, dass es eine Beziehung giebt, welche die unter den Begriff H fallenden Gegenstände den unter den Begriff G fallenden beiderseits eindeutig zuordnet, wenn es eine Beziehung ϕ giebt, welche die unter den Begriff F fallenden Gegenstände den unter den Begriff G fallenden beiderseits eindeutig zuordnet, und wenn es eine Beziehung ψ giebt, welche die unter den Begriff H fallenden Gegenstände den unter den Begriff F fallenden beiderseits eindeutig zuordnet. Folgende Anordnung der Buchstaben wird dies übersichtlicher machen:

$$H \,\psi\, F \,\phi\, G.$$

Eine solche Beziehung kann in der That angegeben werden: sie liegt in dem Inhalte

„es giebt einen Gegenstand, zu dem c in der Beziehung ψ steht, und der zu b in der Beziehung ϕ steht,"

wenn wir davon c und b absondern (als Beziehungspunkte betrachten). Man kann zeigen, dass diese Beziehung eine beiderseits eindeutige ist, und dass sie die unter den Begriff H fallenden Gegenstände den unter den Begriff G fallenden zuordnet.

In ähnlicher Weise kann auch der andere Satz bewiesen werden*). Diese Andeutungen werden hoffentlich genügend erkennen lassen, dass wir hierbei keinen Beweisgrund der Anschauung zu entnehmen brauchen, und dass sich mit unsern Definitionen etwas machen lässt.

§ 74. Wir können nun zu den Erklärungen der einzelnen Zahlen übergehn.

---

*) Desgleichen die Umkehrung: Wenn die Zahl, welche dem Begriffe F zukommt, dieselbe ist wie die, welche dem Begriffe G zukommt, so ist der Begriff F dem Begriffe G gleichzahlig.

then it is also equal to the concept G;

and

if the concept H is equal to the concept G,
then it is also equal to the concept F.

The first proposition amounts to this, that there exists a relation which correlates one to one the objects falling under the concept H with those falling under the concept G, if there exists a relation ϕ which correlates one to one the objects falling under the concept F with those falling under the concept G and if there exists also a relation ψ which correlates one to one the objects falling under the concept H with those falling under the concept F. The following arrangement of letters will make this easier to grasp:

$$H\,\psi\,F\,\phi\,G.$$

Such a relation can in fact be given: it is to be found in the judgement-content

"there exists an object to which c stands in the relation ψ and which stands to b in the relation ϕ",

if we subtract from it c and b—take them, that is, as the terms of the relation. It can be shown that this relation is one-one, and that it correlates the objects falling under the concept H with those falling under the concept G.

A similar proof can be given of the second proposition also.[1] And with that, I hope, enough has been indicated of my methods to show that our proofs are not dependent at any point on borrowings from intuition, and that our definitions can be used to some purpose.

§ 74. We can now pass on to the definitions of the individual numbers.

---

[1] And likewise of the converse: If the number which belongs to the concept F is the same as that which belongs to the concept G, then the concept F is equal to the concept G.

87

Weil unter den Begriff „sich selbst ungleich" nichts fällt, erkläre ich:

o ist die Anzahl, welche dem Begriffe „sich selbst ungleich" zukommt.

Vielleicht nimmt man daran Anstoss, dass ich hier von einem Begriffe spreche. Man wendet vielleicht ein, dass ein Widerspruch darin enthalten sei, und errinert an die alten Bekannten das hölzerne Eisen und den viereckigen Kreis. Nun ich meine, dass die gar nicht so schlimm sind, wie sie gemacht werden. Zwar nützlich werden sie grad nicht sein; aber schaden können sie auch nichts, wenn man nur nicht voraussetzt, dass etwas unter sie falle; und das thut man durch den blossen Gebrauch der Begriffe noch nicht. Dass ein Begriff einen Widerspruch enthalte, ist nicht immer so offensichtlich, dass es keiner Untersuchung bedürfte; dazu muss man ihn erst haben und logisch ebenso wie jeden andern behandeln. Alles was von Seiten der Logik und für die Strenge der Beweisführung von einem Begriffe verlangt werden kann, ist seine scharfe Begrenzung, dass für jeden Gegenstand bestimmt sei, ob er unter ihn falle oder nicht. Dieser Anforderung genügen nun die einen Widerspruch enthaltenden Begriffe wie „sich selbst ungleich" durchaus; denn man weiss von jedem Gegenstande, dass er nicht unter einen solchen fällt*).

Ich brauche das Wort „Begriff" in der Weise, dass

„a fällt unter den Begriff F"

die allgemeine Form eines beurtheilbaren Inhalts ist, der

---

*) Ganz davon verschieden ist die Definition eines Gegenstandes aus einem Begriffe, unter den er fällt. Der Ausdruck „der grösste ächte Bruch" hat z. B. keinen Inhalt, weil der bestimmte Artikel den Anspruch erhebt, auf einen bestimmten Gegenstand hinzuweisen. Dagegen ist der Begriff „Bruch, der kleiner als 1 und so beschaffen ist, dass kein Bruch, der kleiner als 1 ist, ihn an Grösse übertrifft" ganz unbedenklich, und um beweisen zu können, dass es keinen solchen Bruch gebe, braucht man sogar diesen Begriff, obgleich er einen Widerspruch enthält. Wenn man aber durch

Since nothing falls under the concept "not identical with itself", I define nought as follows:

> o is the Number which belongs to the concept "not identical with itself".

Some may find it shocking that I should speak of a concept in this connexion. They will object, very likely, that it contains a contradiction and is reminiscent of our old friends the square circle and wooden iron. Now I believe that these old friends are not so black as they are painted. To be of any use is, I admit, the last thing we should expect of them; but at the same time, they cannot do any harm, if only we do not assume that there is anything which falls under them—and to that we are not committed by merely using them. That a concept contains a contradiction is not always obvious without investigation; but to investigate it we must first possess it and, in logic, treat it just like any other. All that can be demanded of a concept from the point of view of logic and with an eye to rigour of proof is only that the limits to its application should be sharp, that it should be determined, with regard to every object whether it falls under that concept or not. But this demand is completely satisfied by concepts which, like "not identical with itself", contain a contradiction; for of every object we know that it does not fall under any such concept.[1]

On my use of the word "concept",

> "*a* falls under the concept F"

is the general form of a judgement-content which deals with

---

[1] The definition of an object in terms of a concept under which it falls is a very different matter. For example, the expression "the largest proper fraction" has no content, since the definite article claims to refer to a definite object. On the other hand, the concept "fraction smaller than 1 and such that no fraction smaller than one exceeds it in magnitude" is quite unexceptionable: in order, indeed, to prove that there exists no such fraction, we must make use of just this concept, despite its containing a contradiction. If, however, we wished to

von einem Gegenstande a handelt und der beurtheilbar bleibt, was man auch für a setze. Und in diesem Sinne ist

„a fällt unter den Begriff „ „ sich selbst ungleich" " "

gleichbedeutend mit

„a ist sich selbst ungleich"

oder

„a ist nicht gleich a".

Ich hätte zur Definition der o jeden andern Begriff nehmen können, unter den nichts fällt. Es kam mir aber darauf an, einen solchen zu wählen, von dem dies rein logisch bewiesen werden kann; und dazu bietet sich am bequemsten „sich selbst ungleich" dar, wobei ich für „gleich" die vorhin angeführte Erklärung Leibnizens gelten lasse, die rein logisch ist.

§ 75. Es muss sich nun mittels der früheren Festsetzungen beweisen lassen, dass jeder Begriff, unter den nichts fällt, gleichzahlig mit jedem Begriffe ist, unter den nichts fällt, und nur mit einem solchen, woraus folgt, dass o die Anzahl ist, welche einem solchen Begriffe zukommt, und dass kein Gegenstand unter einen Begriff fällt, wenn die Zahl, welche diesem zukommt, die o ist.

Nehmen wir an, weder unter den Begriff F noch unter den Begriff G falle ein Gegenstand, so haben wir, um die Gleichzahligkeit zu beweisen, eine Beziehung ϕ nöthig, von der die Sätze gelten:

jeder Gegenstand, der unter F fällt, steht in der Beziehung ϕ zu einem Gegenstande, der unter G fällt; zu jedem Gegenstande, der unter G fällt, steht ein unter F fallender in der Beziehung ϕ.

diesen Begriff einen Gegenstand bestimmen wollte, der unter ihn fällt, wäre es allerdings nöthig, zweierlei vorher zu zeigen:

1. dass unter diesen Begriff ein Gegenstand falle;
2. dass nur ein einziger Gegenstand unter ihn falle.

Da nun schon der erste dieser Sätze falsch ist, so ist der Ausdruck „der grösste ächte Bruch" sinnlos.

an object *a* and permits of the insertion for *a* of anything whatever. And in this sense

"*a* falls under the concept 'not identical with itself' "

has the same meaning as

"*a* is not identical with itself"

or

"*a* is not identical with *a*".

I could have used for the definition of nought any other concept under which no object falls. But I have made a point of choosing one which can be proved to be such on purely logical grounds; and for this purpose "not identical with itself" is the most convenient that offers, taking for the definition of "identical" the one from LEIBNIZ given above [(§ 65)], which is in purely logical terms.

§ 75. Now it must be possible to prove, by means of what has already been laid down, that every concept under which no object falls is equal to every other concept under which no object falls, and to them alone; from which it follows that o is the Number which belongs to any such concept, and that no object falls under any concept if the number which belongs to that concept is o.

If we assume that no object falls under either the concept *F* or the concept *G*, then in order to prove them equal we have to find a relation ϕ which satisfies the following conditions:

> every object which falls under *F* stands in the relation ϕ to an object which falls under *G*; and to every object which falls under *G* there stands in the relation ϕ an object falling under *F*.

use this concept for defining an object falling under it, it would, of course, be necessary first to show two distinct things:

1. that some object falls under this concept;
2. that only one object falls under it.

Now since the first of these propositions, not to mention the second, is false, it follows that the expression "the largest proper fraction" is senseless.

89

Nach dem, was früher über die Bedeutung dieser Ausdrücke gesagt ist, erfüllt bei unsern Voraussetzungen jede Beziehung diese Bedingungen, also auch die Gleichheit, die obendrein beiderseits eindeutig ist; denn es gelten die beiden oben dafür verlangten Sätze.

Wenn dagegen unter G ein Gegenstand fällt z. B. a, während unter F keiner fällt so bestehen die beiden Sätze

„a fällt unter G"

und

„kein unter F fallender Gegenstand steht zu a in

der Beziehung ϕ"

mit einander für jede Beziehung ϕ; denn der erste ist nach der ersten Voraussetzung richtig und der zweite nach der zweiten. Wenn es nämlich keinen unter F fallenden Gegenstand giebt, so giebt es auch keinen solchen, der in irgendeiner Beziehung zu a stände. Es giebt also keine Beziehung, welche nach unserer Erklärung die unter F den unter G fallenden Gegenständen zuordnete, und demnach sind die Begriffe F und G ungleichzahlig.

§ 76. Ich will nun die Beziehung erklären, in der je zwei benachbarte Glieder der natürlichen Zahlenreihe zu einander stehen. Der Satz:

„es giebt einen Begriff F und einen unter ihn fallenden Gegenstand x der Art, dass die Anzahl, welche dem Begriffe F zukommt, n ist, und dass die Anzahl, welche dem Begriffe „ „unter F fallend aber nicht gleich x" " zukommt, m ist"

sei gleichbedeutend mit

„n folgt in der natürlichen Zahlenreihe unmittelbar auf m."

Ich vermeide den Ausdruck „n ist die auf m nächstfolgende Anzahl," weil zur Rechtfertigung des bestimmten Artikels erst zwei Sätze bewiesen werden müssten*). Aus dem-

*) Siehe Anm. auf S. 87 u. 88.

In view of what has been said above [(§ 71)] on the meaning of these expressions, it follows, on our assumption [that no object falls under either concept], that these conditions are satisfied by every relation whatsoever, and therefore among others by identity, which is moreover a one-one relation; for it meets both the requirements laid down [in § 72] above.

If, to take the other case, some object, say a, does fall under G, but still none falls under F, then the two propositions

"a falls under G"

and

"no object falling under F stands to a in the relation ϕ"

are both true together for every relation ϕ; for the first is made true by our first assumption and the second by our second assumption. If, that is, there exists no object falling under F, then a fortiori there exists no object falling under F which stands to a in any relation whatsoever. There exists, therefore, no relation by which the objects falling under F can be correlated with those falling under G so as to satisfy our definition [of equality], and accordingly the concepts F and G are unequal.

§ 76. I now propose to define the relation in which every two adjacent members of the series of natural numbers stand to each other. The proposition:

"there exists a concept F, and an object falling under it x, such that the Number which belongs to the concept F is n and the Number which belongs to the concept 'falling under F but not identical with x' is m"

is to mean the same as

"n follows in the series of natural numbers directly after m".

I avoid the expression "n is *the* Number following next after m", because the use of the definite article cannot be justified until we have first proved two propositions.[1] For

---

[1] See note on p. 87ᵉ f.

selben Grunde sage ich hier noch nicht „n = m + 1“; denn auch durch das Gleichheitszeichen wird (m + 1) als Gegenstand bezeichnet.

§ 77. Un nun auf die Zahl 1 zu kommen, müssen wir zunächst zeigen, dass es etwas giebt, was in der natürlichen Zahlenreihe unmittelbar auf o folgt.

Betrachten wir den Begriff — oder, wenn man lieber will, das Prädicat — „gleich o“! Unter diesen fällt die o. Unter den Begriff „gleich o aber nicht gleich o“ fällt dagegen kein Gegenstand, sodass o die Anzahl ist, welche diesem Begriffe zukommt. Wir haben demnach einen Begriff „gleich o“ und einen unter ihn fallenden Gegenstand o, von denen gilt:

die Anzahl, welche dem Begriffe „gleich o“ zukommt, ist gleich der Anzahl, welche dem Begriffe „gleich o“ zukommt;

die Anzahl, welche dem Begriffe „gleich o aber nicht gleich o“ zukommt, ist die o.

Also folgt nach unserer Erklärung die Anzahl, welche dem Begriffe „gleich o“ zukommt, in der natürlichen Zahlenreihe unmittelbar auf o.

Wenn wir nun definiren:

1 ist die Anzahl, welche dem Begriffe „gleich o“ zukommt,

so können wir den letzten Satz so ausdrücken:

1 folgt in der natürlichen Zahlenreihe unmittelbar auf o.

Es ist vielleicht nicht überflussig zu bemerken, dass die Definition der 1 zu ihrer objectiven Rechtmässigkeit keine beobachtete Thatsache*) voraussetzt; denn man verwechselt leicht damit, dass gewisse subjective Bedingungen erfüllt sein müssen, um uns die Definition möglich zu machen, und dass uns Sinneswahrnehmungen dazu veranlassen**).

---

*) Satz ohne Allgemeinheit.
**) Vergl. B. Erdmann, die Axiome der Geometrie S. 164.

the same reason I do not yet say at this point "$n = m + 1$," for to use the symbol $=$ is likewise to designate $(m + 1)$ an object.

§ 77. Now in order to arrive at the number 1, we have first of all to show that there is something which follows in the series of natural numbers directly after 0.

Let us consider the concept—or, if you prefer it, the predicate—"identical with 0". Under this falls the number 0. But under the concept "identical with 0 but not identical with 0", on the other hand, no object falls, so that 0 is the Number which belongs to this concept. We have, therefore, a concept "identical with 0" and an object falling under it 0, of which the following propositions hold true:

the Number which belongs to the concept "identical with 0" is identical with the Number which belongs to the concept "identical with 0";

the Number which belongs to the concept "identical with 0 but not identical with 0" is 0.

Therefore, on our definition [(§ 76)], the Number which belongs to the concept "identical with 0" follows in the series of natural numbers directly after 0.

Now if we give the following definition:

1 is the Number which belongs to the concept "identical with 0",

we can then put the preceding conclusion thus:

1 follows in the series of natural numbers directly after 0.

It is perhaps worth pointing out that our definition of the number 1 does not presuppose, for its objective legitimacy, any matter of observed fact.[1] It is easy to get confused over this, seeing that certain subjective conditions must be satisfied if we are to be able to arrive at the definition, and that sense experiences are what prompt us to frame it.[2] All this, how-

---

[1] Non-general proposition.
[2] Cf. B. Erdmann, *Die Axiome der Geometrie*, p. 164.

Dies kann immerhin zutreffen, ohne dass die abgeleiteten Sätze aufhören, a priori zu sein. Zu solchen Bedingungen gehört z. B. auch, dass Blut in hinreichender Fülle und richtiger Beschaffenheit das Gehirn durchströme — wenigstens soviel wir wissen; — aber die Wahrheit unseres letzten Satzes ist davon unabhängig; sie bleibt bestehen, auch wenn dies nicht mehr stattfindet; und selbst, wenn alle Vernunftwesen einmal gleichzeitig in einen Winterschlaf verfallen sollten, so würde sie nicht etwa so lange aufgehoben sein, sondern ganz ungestört bleiben. Die Wahrheit eines Satzes ist eben nicht sein Gedachtwerden.

§ 78. Ich lasse hier einige Sätze folgen, die mittels unserer Definitionen zu beweisen sind. Der Leser wird leicht übersehen, wie dies geschehen kann.

1. Wenn a in der natürlichen Zahlenreihe unmittelbar auf o folgt, so ist a = 1.

2. Wenn 1 die Anzahl ist, welche einem Begriffe zukommt, so giebt es einen Gegenstand, der unter den Begriff fällt.

3. Wenn 1 die Anzahl ist, welche einem Begriffe F zukommt; wenn der Gegenstand x unter den Begriff F fällt, und wenn y unter den Begriff F fällt, so ist x = y; d. h. x ist dasselbe wie y.

4. Wenn unter einen Begriff F ein Gegenstand fällt, und wenn allgemein daraus, dass x unter den Begriff F fällt, und dass y unter den Begriff F fällt, geschlossen werden kann, dass x = y ist, so ist 1 die Anzahl, welche dem Begriffe F zukommt.

5. Die Beziehung von m zu n, die durch den Satz:

    „n folgt in der natürlichen Zahlenreihe unmittelbar auf m"

gesetzt wird, ist eine beiderseits eindeutige.

Hiermit ist noch nicht gesagt, dass es zu jeder Anzahl eine andere gebe, welche auf sie oder auf welche sie in der Zahlenreihe unmittelbar folge.

ever, may be perfectly correct, without the propositions so arrived at ceasing to be a priori. One such condition is, for example, that blood of the right quality must circulate in the brain in sufficient volume—at least so far as we know; but the truth of our last proposition does not depend on this; it still holds, even if the circulation stops; and even if all rational beings were to take to hibernating and fall asleep simultaneously, our proposition would not be, say, cancelled for the duration, but would remain quite unaffected. For a proposition to be true is just not the same thing as for it to be thought.

§ 78. I proceed to give here a list of several propositions to be proved by means of our definitions. The reader will easily see for himself in outline how this can be done.

1. If a follows in the series of natural numbers directly after o, then a is $= 1$.

2. If 1 is the Number which belongs to a concept, then there exists an object which falls under that concept.

3. If 1 is the Number which belongs to a concept F; then, if the object x falls under the concept F and if y falls under the concept F, x is $= y$; that is, x is the same as y.

4. If an object falls under the concept F, and if it can be inferred generally from the propositions that x falls under the concept F and that y falls under the concept F that x is $= y$, then 1 is the Number which belongs to the concept F.

5. The relation of m to n which is established by the proposition:
   "n follows in the series of natural numbers directly after m"
   is a one-one relation.

There is nothing in this so far to state that for every Number there exists another Number which follows directly after it, or after which it directly follows, in the series of natural numbers.

92

6. Jede Anzahl ausser der o folgt in der natürlichen Zahlenreihe unmittelbar auf eine Anzahl.

§ 79. Um nun beweisen zu können, dass auf jede Anzahl (n) in der natürlichen Zahlenreihe eine Anzahl unmittelbar folge, muss man einen Begriff aufweisen, dem diese letzte Anzahl zukommt. Wir wählen als diesen

„der mit n endenden natürlichen Zahlenreihe ange-
> hörend",

der zunächst erklärt werden muss.

Ich wiederhole zunächst mit etwas andern Worten die Definition, welche ich in meiner „Begriffsschrift" vom Folgen in einer Reihe gegeben habe.

Der Satz

> „wenn jeder Gegenstand, zu dem x in der Beziehung ϕ steht, unter den Begriff F fällt, und wenn daraus, dass d unter den Begriff F fällt, allgemein, was auch d sei, folgt, dass jeder Gegenstand, zu dem d in der Beziehung ϕ steht, unter den Begriff F fälle, so fällt y unter den Begriff F, was auch F für ein Begriff sein möge"

sei gleichbedeutend mit

> „y folgt in der ϕ - Reihe auf x"

und mit

> „x geht in der ϕ - Reihe dem y vorher."

§ 80. Einige Bemerkungen hierzu werden nicht überflüssig sein. Da die Beziehung ϕ unbestimmt gelassen ist, so ist die Reihe nicht nothwendig in der Form einer räumlichen und zeitlichen Anordnung zu denken, obwohl diese Fälle nicht ausgeschlossen sind.

Man könnte vielleicht eine andere Erklärung für natürlicher halten z. B.: wenn man von x ausgehend seine Aufmerksamkeit immer von einem Gegenstande zu einem andern lenkt, zu welchem er in der Beziehung ϕ steht, und wenn man auf diese Weise schliesslich y erreichen kann, so sagt man y folge in der ϕ - Reihe auf x.

6. Every Number except o follows in the series of natural numbers directly after a Number.

§ 79. Now in order to prove that after every Number (*n*) in the series of natural numbers a Number directly follows, we must produce a concept to which this latter Number belongs. For this we shall choose the concept

"member of the series of natural numbers ending with *n*",

which requires first to be defined.

To start with, let me repeat in slightly different words the definition of following in a series given in my *Begriffs-schrift* [*Concept Writing*]*:

The proposition

'if every object to which *x* stands in the relation φ falls under the concept *F*, and if from the proposition that *d* falls under the concept *F* it follows universally, whatever *d* may be, that every object to which *d* stands in the relation φ falls under the concept *F*, then *y* falls under the concept *F*, whatever concept *F* may be"

is to mean the same as

"*y* follows in the φ-series after *x*"

and again the same as

"*x* comes in the φ-series before *y*".

§ 80. It will not be time wasted to make a few comments on this. First, since the relation φ has been left indefinite, the series is not necessarily to be conceived in the form of a spatial and temporal arrangement, although these cases are not excluded.

Next, there may be those who will prefer some other definition as being more natural, as for example the following: if starting from *x* we transfer our attention continually from one object to another to which it stands in the relation φ, and if by this procedure we can finally reach *y*, then we say that *y* follows in the φ-series after *x*.

* [Cp. § 91 and notes.]

93

Dies ist eine Weise die Sache zu untersuchen, keine Definition. Ob wir bei der Wanderung unserer Aufmerksamkeit y erreichen, kann von mancherlei subjectiven Nebenumständen abhangen z. B. von der uns zu Gebote stehenden Zeit, oder von unserer Kenntniss der Dinge. Ob y auf x in der ϕ - Reihe folgt, hat im Allgemeinen gar nichts mit unserer Aufmerksamkeit und den Bedingungen ihrer Fortbewegung zu thun, sondern ist etwas Sachliches, ebenso wie ein grünes Blatt gewisse Lichtstrahlen reflectirt, mögen sie nun in mein Auge fallen und Empfindung hervorrufen oder nicht, ebenso wie ein Salzkorn in Wasser löslich ist, mag ich es ins Wasser werfen und den Vorgang beobachten oder nicht, und wie es selbst dann noch löslich ist, wenn ich gar nicht die Möglichkeit habe, einen Versuch damit anzustellen.

Durch meine Erklärung ist die Sache aus dem Bereiche subjectiver Möglichkeiten in das der objectiven Bestimmtheit erhoben. In der That: dass aus gewissen Sätzen ein anderer folgt, ist etwas Objectives, von den Gesetzen der Bewegung unserer Aufmerksamkeit Unabhängiges, und es ist dafür einerlei, ob wir den Schluss wirklich machen oder nicht. Hier haben wir ein Merkmal, das die Frage überall entscheidet, wo sie gestellt werden kann, mögen wir auch im einzelnen Falle durch äussere Schwierigkeiten verhindert sein, zu beurtheilen, ob es zutrifft. Das ist für die Sache selbst gleichgiltig.

Wir brauchen nicht immer alle Zwischenglieder vom Anfangsgliede bis zu einem Gegenstande zu durchlaufen, um gewiss zu sein, dass er auf jenes folgt. Wenn z. B. gegeben ist, dass in der ϕ - Reihe b auf a und c auf b folgt, so können wir nach unserer Erklärung schliessen, dass c auf a folgt, ohne die Zwischenglieder auch nur zu kennen.

Durch diese Definition des Folgens in einer Reihe wird es allein möglich, die Schlussweise von n auf (n + 1), welche scheinbar der Mathematik eigenthümlich ist, auf die allgemeinen logischen Gesetze zurückzuführen.

Now this describes a way of discovering that y follows, it does not define what is meant by y's following. Whether, as our attention shifts, we reach y may depend on all sorts of subjective contributory factors, for example on the amount of time at our disposal or on the extent of our familiarity with the things concerned. Whether y follows in the ϕ-series after x has in general absolutely nothing to do with our attention and the circumstances in which we transfer it; on the contrary, it is a question of fact, just as much as it is a fact that a green leaf reflects light rays of certain wave-lengths whether or not these fall into my eye and give rise to a sensation, and a fact that a grain of salt is soluble in water whether or not I drop it into water and observe the result, and a further fact that it remains still soluble even when it is utterly impossible for me to make any experiment with it.

My definition lifts the matter onto a new plane; it is no longer a question of what is subjectively possible but of what is objectively definite. For in literal fact, that one proposition follows from certain others is something objective, something independent of the laws that govern the movements of our attention, and something to which it is immaterial whether we actually draw the conclusion or not. What I have provided is a criterion which decides in every case the question Does it follow after?, wherever it can be put; and however much in particular cases we may be prevented by extraneous difficulties from actually reaching a decision, that is irrelevant to the fact itself.

We have no need always to run through all the members of a series intervening between the first member and some given object, in order to ascertain that the latter does follow after the former. Given, for example, that in the ϕ-series b follows after a and c after b, then we can deduce from our definition that c follows after a, without even knowing the intervening members of the series.

Only by means of this definition of following in a series is it possible to reduce the argument from n to $(n + 1)$, which on the face of it is peculiar to mathematics, to the general laws of logic.

§ 81. Wenn wir nun als Beziehung ϕ diejenige haben, in welche m zu n gesetzt wird durch den Satz

„n folgt in der natürlichen Zahlenreihe unmittelbar auf m,"

so sagen wir statt „ϕ - Reihe" „natürliche Zahlenreihe".
Ich definire weiter:

der Satz

„y folgt in der ϕ - Reihe auf x oder y ist dasselbe wie x"

sei gleichbedeutend mit

„y gehört der mit x anfangenden ϕ - Reihe an"

und mit

„x gehört der mit y endenden ϕ - Reihe an".

Demnach gehört a der mit n endenden natürlichen Zahlenreihe an, wenn n entweder in der natürlichen Zahlenreihe auf a folgt oder gleich a ist*).

§ 82. Es ist nun zu zeigen, dass — unter einer noch anzugebenden Bedingung — die Anzahl, welche dem Begriffe

„der mit n endenden natürlichen Zahlenreihe angehörend"

zukommt, auf n in der natürlichen Zahlenreihe unmittelbar folgt. Und damit ist dann bewiesen, dass es eine Anzahl giebt, welche auf n in der natürlichen Zahlenreihe unmittelbar folgt, dass es kein letztes Glied dieser Reihe giebt. Offenbar kann dieser Satz auf empirischen Wege oder durch Induction nicht begründet werden.

Es würde hier zu weit führen, den Beweis selbst zu geben. Nur sein Gang mag kurz angedeutet werden. Es ist zu beweisen

1. wenn a in der natürlichen Zahlenreihe unmittelbar au d folgt, und wenn von d gilt:

---

*) Wenn n keine Anzahl ist, so gehört nur n selbst der mit n endenden natürlichen Zahlenreihe an. Man stosse sich nicht an dem Ausdrucke!

§ 81. If now we have for our relation ϕ the relation of m to n established by the proposition

"n follows in the series of natural numbers directly after m,"

then we shall say instead of "ϕ-series" "series of natural numbers".

I add the following further definition:

The proposition

"y follows in the ϕ-series after x or y is the same as x"

is to mean the same as

"y is a member of the ϕ-series beginning with x"

and again the same as

"x is a member of the ϕ-series ending with y".

It follows that a is a member of the series of natural numbers ending with n, if n either follows in the series of natural numbers after a or is identical with a.[1]

§ 82. It is now to be shown that—subject to a condition still to be specified—the Number which belongs to the concept

"member of the series of natural numbers ending with n"

follows in the series of natural numbers directly after n. And in thus proving that there exists a Number which follows in the series of natural numbers directly after n, we shall have proved at the same time that there is no last member of this series. Obviously this proposition cannot be established on empirical lines or by induction.

To give the proof in full here would take us too far afield. I can only indicate briefly the way it goes. It is to be proved that

1. if a follows in the series of natural numbers directly after d, and if it is true of d that:

---

[1] If n is not a Number, then n itself is the only member of the series of natural numbers ending with n,—if that is not too shocking a way of putting it.

die Anzahl, welche dem Begriffe

„der mit d endenden natürlichen Zahlenreihe angehörend"

zukommt, folgt in der natürlichen Zahlenreihe unmittelbar auf d,

so gilt auch von a:

die Anzahl, welche dem Begriffe

„der mit a endenden natürlichen Zahlenreihe angehörend"

zukommt, folgt in der natürlichen Zahlenreihe unmittelbar auf a.

Es ist zweitens zu beweisen, dass von der o das gilt,
was in den eben ausgesprochenen Sätzen von d und von a
ausgesagt ist, und dann zu folgern, dass es auch von n gilt,
wenn n der mit o anfangenden natürlichen Zahlenreihe angehört. Diese Schlussweise ist eine Anwendung der Definition,
die ich von dem Ausdrucke

„y folgt in der natürlichen Zahlenreihe auf x"

gegeben habe, indem man als Begriff F jene gemeinsame
Aussage von d und von a, von o und von n zu nehmen hat.

§ 83. Um den Satz (1) des vorigen § zu beweisen,
müssen wir zeigen, dass a die Anzahl ist, welche dem Begriffe „der mit a endenden natürlichen Zahlenreihe angehörend, aber nicht gleich a" zukommt. Und dazu ist wieder
zu beweisen, dass dieser Begriff gleichen Umfanges mit dem
Begriffe „der mit d endenden natürlichen Zahlenreihe angehörend" ist. Hierfür bedarf man des Satzes, dass kein
Gegenstand, welcher der mit o anfangenden natürlichen
Zahlenreihe angehört, auf sich selbst in der natürlichen
Zahlenreihe folgen kann. Dies muss ebenfalls mittels unserer
Definition des Folgens in einer Reihe, wie oben angedeutet
ist, bewiesen werden*).

*) E. Schröder scheint a. a. O. S. 63 diesen Satz als Folge einer
auch anders denkbaren Bezeichnungsweise anzusehen. Es macht sich
auch hier der Uebelstand bemerkbar, der seine ganze Darstellung dieser

the Number which belongs to the concept

"member of the series of natural numbers ending with d"

follows in the series of natural numbers directly after d,
then it is also true of a that:

the Number which belongs to the concept

"member of the series of natural numbers ending with a"

follows in the series of natural numbers directly after a.

It is then to be proved, secondly, that what is asserted of d and of a in the propositions just stated holds for the number o. And finally it is to be deduced that it also holds for n if n is a member of the series of natural numbers beginning with o. The argument here is an application of the definition I have given [(§§ 79, 81)] of the expression

"y follows in the series of natural numbers after x",

taking for our concept F what is asserted above [in 1.] of d and a conjointly, but with o and n substituted for d and a.

§ 83. In order to prove the proposition 1. of the last paragraph, we must show that a is the Number which belongs to the concept "member of the series of natural numbers ending with a, but not identical with a". And for this, again, it is necessary to prove that this concept has an extension identical with that of the concept "member of the series of natural numbers ending with d". For this we need the proposition that no object which is a member of the series of natural numbers beginning with o can follow in the series of natural numbers after itself. And this must once again be proved by means of our definition of following in a series, on the lines indicated above.[1]

---

[1] E. Schröder (op. cit., p. 63) seems to regard this proposition as a consequence of a system of notation which could conceivably be different. Here once more we must be struck by the drawback which vitiates his whole treat-

96

Wir werden hierdurch genöthigt, dem Satze, dass die Anzahl, welche dem Begriffe

> „der mit n endenden natürlichen Zahlenreihe ange-
> hörend"

zukommt, in der natürlichen Zahlenreihe unmittelbar auf n folgt, die Bedingung hinzuzufügen, dass n der mit o anfangenden natürlichen Zahlenreihe angehöre. Hierfür ist eine kürzere Ausdrucksweise gebräuchlich, die ich nun erkläre:

> der Satz
> „n gehört der mit o anfangenden natürlichen Zahlen-
> reihe an"

sei gleichbedeutend mit

> „n ist eine endliche Anzahl".

Dann können wir den letzten Satz so ausdrücken: keine endliche Anzahl folgt in der natürlichen Zahlenreihe auf sich selber.

Unendliche Anzahlen.

§ 84. Den endlichen gegenüber stehen die unendlichen Anzahlen. Die Anzahl, welche dem Begriffe „endliche Anzahl" zukommt, ist eine unendliche. Bezeichnen wir sie etwa durch ∞_1! Wäre sie eine endliche, so könnte sie nicht auf sich selber in der natürlichen Zahlenreihe folgen. Man kann aber zeigen, dass ∞_1 das thut.

In der so erklärten unendlichen Anzahl ∞_1 liegt nichts irgendwie Geheimnissvolles oder Wunderbares. „Die Anzahl, welche dem Begriffe F zukommt, ist ∞_1" heisst nun nichts mehr und nichts weniger als: es giebt eine Beziehung, welche die unter den Begriff F fallenden Gegenstände den endlichen

---

Sache beeinträchtigt, dass man nicht recht weiss, ob die Zahl ein Zeichen ist, und was dann dessen Bedeutung, oder ob sie eben diese Bedeutung ist. Daraus, dass man verschiedene Zeichen festsetzt, sodass nie dasselbe wiederkehrt, folgt noch nicht, dass diese Zeichen auch Verschiedenes bedeuten.

It is this that obliges us to attach a condition to the proposition that the Number which belongs to the concept

"member of the series of natural numbers ending with n"

follows in the series of natural numbers directly after n,—the condition, namely, that n must be a member of the series of natural numbers beginning with o. For this there is a convenient abbreviation, which I define as follows:

the proposition

"n is a member of the series of natural numbers beginning with o"

is to mean the same as

"n is a finite Number".

We can thus formulate the last proposition above as follows: no finite Number follows in the series of natural numbers after itself.

Infinite Numbers.

§ 84. Contrasted with the finite Numbers are the infinite Numbers. The Number which belongs to the concept "finite Number" is an infinite Number. Let us symbolize it by, say, ∞_1. If it were a finite Number, it could not follow in the series of natural numbers after itself. But it can be shown that this is what ∞_1 does.

About the infinite Number ∞_1 so defined there is nothing mysterious or wonderful. "The Number which belongs to the concept F is ∞_1" means no more and no less than this: that there exists a relation which correlates one to one the objects falling under the concept F with the finite Numbers. In terms

ment of this matter,—that we do not really know whether the number is a symbol and if so what its meaning is, or whether the number itself is the meaning of the symbol. He is not entitled to infer, from the fact that we arrange for our symbols to differ so that the same one never recurs, that the meanings of those symbols are therefore also different.

Anzahlen beiderseits eindeutig zuordnet. Dies ist nach unseren Erklärungen ein ganz klarer und unzweideutiger Sinn; und das genügt, um den Gebrauch des Zeichens ∞_1 zu rechtfertigen und ihm eine Bedeutung zu sichern. Dass wir uns keine Vorstellung von einer unendlichen Anzahl bilden können, ist ganz unerheblich und würde endliche Anzahlen ebenso treffen. Unsere Anzahl ∞_1 hat auf diese Weise etwas ebenso Bestimmtes wie irgendeine endliche: sie ist zweifellos als dieselbe wiederzuerkennen und von einer andern zu unterscheiden.

§ 85. Vor Kurzem hat G. Cantor in einer bemerkenswerthen Schrift*) unendliche Anzahlen eingeführt. Ich stimme ihm durchaus in der Würdigung der Ansicht bei, welche überhaupt nur die endlichen Anzahlen als wirklich gelten lassen will. Sinnlich wahrnehmbar und räumlich sind weder diese noch die Brüche, noch die negativen, irrationalen und complexen Zahlen; und wenn man wirklich nennt, was auf die Sinne wirkt, oder was wenigstens Wirkungen hat, die Sinneswahrnehmungen zur nähern oder entferntern Folge haben können, so ist freilich keine dieser Zahlen wirklich. Aber wir brauchen auch solche Wahrnehmungen gar nicht als Beweisgründe für unsere Lehrsätze. Einen Namen oder ein Zeichen, das logisch einwurfsfrei eingeführt ist, können wir in unsern Untersuchungen ohne Scheu gebrauchen, und so ist unsere Anzahl ∞_1 so gerechtfertigt wie die Zwei oder die Drei.

Indem ich hierin, wie ich glaube, mit Cantor übereinstimme, weiche ich doch in der Benennung etwas von ihm ab. Meine Anzahl nennt er „Mächtigkeit," während sein Begriff**) der Anzahl auf die Anordnung Bezug nimmt. Für

---

*) Grundlagen einer allgemeinen Mannichfaltigkeitslehre. Leipzig, 1883.

**) Dieser Ausdruck kann der früher hervorgehobenen Objectivität des Begriffes zu widersprechen scheinen; aber subjectiv ist hier nur die Benennung.

of our definitions this has a perfectly clear and unambiguous sense; and that is enough to justify the use of the symbol ∞ and to assure it of a meaning. That we cannot form any idea of an infinite Number is of absolutely no importance; the same is equally true of finite Numbers. So regarded, our Number ∞_1, has a character as definite as that of any finite Number; it can be recognized again beyond doubt as the same, and can be distinguished from every other.

§ 85. It is only recently that infinite Numbers have been introduced, in a remarkable work by G. CANTOR.[1] I heartily share his contempt for the view that in principle only finite Numbers ought to be admitted as actual. Perceptible by the senses these are not, nor are they spatial—any more than fractions are, or negative numbers, or irrational or complex numbers; and if we restrict the actual to what acts on our senses or at least produces effects which may cause sense-perceptions as near or remote consequences, then naturally no number of any of these kinds is actual. But it is also true that we have no need at all to appeal to any such sense-perceptions in proving our theorems. Any name or symbol that has been introduced in a logically unexceptionable manner can be used in our enquiries without hesitation, and here our Number ∞_1 is as sound as 2 or 3.

While in this I agree, as I believe, with CANTOR, my terminology diverges to some extent from his. For my Number he uses "power", while his concept[2] of Number has reference to arrangement in an order. Finite Numbers,

---

[1] Op. cit., p. 74^e above.

[2] This expression may seem to conflict with my earlier insistence on the objective nature of concepts; but all that I mean is subjective here is his use of the *word*.

endliche Anzahlen ergiebt sich freilich doch eine Unabhängigkeit von der Reihenfolge, dagegen nicht für unendlich-
grosse. Nun enthält der Sprachgebrauch des Wortes „Anzahl"
und der Frage „wieviele?" keine Hinweisung auf eine
bestimmte Anordnung. Cantors Anzahl antwortet vielmehr
auf die Frage: „das wievielste Glied in der Succession ist
das Endglied?" Darum scheint mir meine Benennung besser
mit dem Sprachgebrauche übereinzustimmen. Wenn man die
Bedeutung eines Wortes erweitert, so wird man darauf zu
achten haben, dass möglichst viele allgemeine Sätze ihre
Geltung behalten und zumal so grundlegende, wie für die
Anzahl die Unabhängigkeit von der Reihenfolge ist. Wir
haben gar keine Erweiterung nöthig gehabt, weil unser
Begriff der Anzahl sofort auch unendliche Zahlen umfasst.

§ 86. Um seine unendlichen Anzahlen zu gewinnen,
führt Cantor den Beziehungsbegriff des Folgens in einer
Succession ein, der von meinem „Folgen in einer Reihe"
abweicht. Nach ihm würde z. B. eine Succession entstehen,
wenn man die endlichen positiven ganzen Zahlen so anordnete,
dass die unpaaren in ihrer natürlichen Reihenfolge für sich
und ebenso die paaren unter sich auf einander folgten, ferner
festgesetzt wäre, dass jede paare auf jede unpaare folgen
solle. In dieser Succession würde z. B. o auf 13 folgen.
Es würde aber keine Zahl unmittelbar der o vorhergehen.
Dies ist nun ein Fall, der in dem von mir definirten Folgen
in der Reihe nicht vorkommen kann. Man kann streng beweisen, ohne ein Axiom der Anschauung zu benutzen, dass
wenn y auf x in der ϕ - Reihe folgt, es einen Gegenstand
giebt, der in dieser Reihe dem y unmittelbar vorhergeht.
Mir scheinen nun genaue Definitionen des Folgens in der
Succession und der cantorschen Anzahl noch zu fehlen. So
beruft sich Cantor auf die etwas geheimnissvolle „innere
Anschauung," wo ein Beweis aus Definitionen anzustreben
und wohl auch möglich wäre. Denn ich glaube vorauszusehen, wie sich jene Begriffe bestimmen liessen. Jedenfalls

certainly, emerge as independent nevertheless of sequence in series, but not so transfinite Numbers. But now in ordinary use the word "Number" and the question "how many?" have no reference to arrangement in a fixed order. CANTOR's Number gives rather the answer to the question: "the how-manyeth member in the succession is the last member?" So that it seems to me that my terminology accords better with ordinary usage. If we extend the meaning of a word, we should take care that, so far as possible, no general proposition is invalidated in the process, especially one so fundamental as that which asserts of Number its independence of sequence in series. For us, because our concept of Number has from the outset covered infinite numbers as well, no extension of its meaning has been necessary at all.

§ 86. To obtain his infinite Numbers CANTOR introduces the relation-concept of following in a succession, which differs from my "following in a series". On his account we should get a succession if, for example, we arranged the finite positive whole numbers in an order such that the odd numbers followed one another just as they do, among themselves, in the series of natural numbers, and similarly the even numbers, but with the further stipulation that every even number was to follow after every odd number. In this succession o, for instance, would follow after 13. But no number would come directly before o. Now this is a situation which cannot arise on my definition of following in the series. It can be strictly proved, without appeal to any axiom borrowed from intuition, that if y follows in the ϕ-series after x then there exists an object which comes in that series directly before y. Now it looks to me as though precise definitions of following in the succession and of Number in CANTOR's sense are still wanting. Thus CANTOR appeals to the rather mysterious "inner intuition", where he ought to have made an effort to find, and indeed could actually have found, a proof from definitions. For I think I can anticipate how his two concepts could have been made precise. At any rate, nothing in what I have said is

will ich durch diese Bemerkungen, deren Berechtigung und Fruchtbarkeit durchaus nicht angreifen. Im Gegentheil begrüsse ich in diesen Untersuchungen eine Erweiterung der Wissenschaft besonders deshalb, weil durch sie ein rein arithmetischer Weg zu höhern unendlichgrossen Anzahlen (Mächtigkeiten) gebahnt ist.

V. Schluss.

§ 87. Ich hoffe in dieser Schrift wahrscheinlich gemacht zu haben, dass die arithmetischen Gesetze analytische Urtheile und folglich a priori sind. Demnach würde die Arithmetik nur eine weiter ausgebildete Logik, jeder arithmetische Satz ein logisches Gesetz, jedoch ein abgeleitetes sein. Die Anwendungen der Arithmetik zur Naturerklärung wären logische Bearbeitungen von beobachteten Thatsachen*); Rechnen wäre Schlussfolgern. Die Zahlgesetze werden nicht, wie Baumann**) meint, eine praktische Bewährung nöthig haben, um in der Aussenwelt anwendbar zu sein; denn in der Aussenwelt, der Gesammtheit des Räumlichen, giebt es keine Begriffe, keine Eigenschaften der Begriffe, keine Zahlen. Also sind die Zahlgesetze nicht eigentlich auf die äussern Dinge anwendbar: sie sind nicht Naturgesetze. Wohl aber sind sie anwendbar auf Urtheile, die von Dingen der Aussenwelt gelten: sie sind Gesetze der Naturgesetze. Sie behaupten nicht einen Zusammenhang zwischen Naturerscheinungen, sondern einen solchen zwischen Urtheilen; und zu diesen gehören auch die Naturgesetze.

§ 88. Kant***) hat den Werth der analytischen Urtheile offenbar — wohl in Folge einer zu engen Begriffsbestimmung — unterschätzt, obgleich ihm der hier benutzte weitere Begriff

---

*) Das Beobachten schliesst selbst schon eine Logische Thätigkeit ein.
**) A. a. O. Bd. II. S. 670.
***) A. a. O. III. S. 39 u. ff.

intended to question in any way their legitimacy or their fertility. On the contrary, I find special reason to welcome in CANTOR's investigations an extension of the frontiers of science, because they have led to the construction of a purely arithmetical route to higher transfinite Numbers (powers).

V. Conclusion.

§ 87. I hope I may claim in the present work to have made it probable that the laws of arithmetic are analytic judgements and consequently a priori. Arithmetic thus becomes simply a development of logic, and every proposition of arithmetic a law of logic, albeit a derivative one. To apply arithmetic in the physical sciences is to bring logic to bear on observed facts;[1] calculation becomes deduction. The laws of number will not, as BAUMANN[2] thinks, need to stand up to practical tests if they are to be applicable to the external world; for in the external world, in the whole of space and all that therein is, there are no concepts, no properties of concepts, no numbers. The laws of number, therefore, are not really applicable to external things; they are not laws of nature. They are, however, applicable to judgements holding good of things in the external world: they are laws of the laws of nature. They assert not connexions between phenomena, but connexions between judgements; and among judgements are included the laws of nature.

§ 88. KANT[3] obviously—as a result, no doubt, of defining them too narrowly—underestimated the value of analytic judgements, though it seems that he did have some inkling

---

[1] Observation itself already includes within it a logical activity.
[2] Op. cit., Vol. II, p. 670.
[3] Op. cit., Vol. III, pp. 39 ff. [Original edns., A6 ff./B10 ff.].

vorgeschwebt zu haben scheint*). Wenn man seine Definition zu Grunde legt, ist die Eintheilung in analytische
und synthetische Urtheile nicht erschöpfend. Er denkt an
den Fall des allgemein bejahenden Urtheils. Dann kann
man von einem Subjectsbegriffe reden und fragen, ob der
Prädicatsbegriff in ihm — zufolge der Definition — enthalten
sei. Wie aber, wenn das Subject ein einzelner Gegenstand
ist? wie, wenn es sich um ein Existentialurtheil handelt?
Dann kann in diesem Sinne gar nicht von einem Subjectsbegriffe die Rede sein. Kant scheint den Begriff durch
beigeordnete Merkmale bestimmt zu denken; das ist aber
eine der am wenigsten fruchtbaren Begriffsbildungen. Wenn
man die oben gegebenen Definitionen überblickt, so wird
man kaum eine von der Art finden. Dasselbe gilt auch von
den wirklich fruchtbaren Definitionen in der Mathematik
z. B. der Stetigkeit einer Function. Wir haben da nicht
eine Reihe beigeordneter Merkmale, sondern eine innigere,
ich möchte sagen organischere Verbindung der Bestimmungen.
Man kann sich den Unterschied durch ein geometrisches
Bild anschaulich machen. Wenn man die Begriffe (oder ihre
Umfänge) durch Bezirke einer Ebene darstellt, so entspricht
dem durch beigeordnete Merkmale definirten Begriffe der
Bezirk, welcher allen Bezirken der Merkmale gemeinsam ist;
er wird durch Theile von deren Begrenzungen umschlossen.
Bei einer solchen Definition handelt es sich also — im Bilde
zu sprechen — darum, die schon gegebenen Linien in neuer
Weise zur Abgrenzung eines Bezirks zu verwenden**). Aber
dabei kommt nichts wesentlich Neues zum Vorschein. Die
fruchtbareren Begriffsbestimmungen ziehen Grenzlinien, die
noch gar nicht gegeben waren. Was sich aus ihnen schliessen
lasse, ist nicht von vornherein zu übersehen; man holt dabei

---

*) S. 43 sagt er, dass ein synthetischer Satz nur dann nach dem Satze
des Widerspruchs eingesehen werden kann, wenn ein andrer synthetischer
Satz vorausgesetzt wird.

**) Ebenso, wenn die Merkmale durch „oder" verbunden sind.

of the wider sense in which I have used the term.[1] On the basis of his definition, the division of judgements into analytic and synthetic is not exhaustive. What he is thinking of is the universal affirmative judgement; there, we can speak of a subject concept and ask—as his definition requires—whether the predicate concept is contained in it or not. But how can we do this, if the subject is an individual object? Or if the judgement is an existential one? In these cases there can simply be no question of a subject concept in KANT's sense. He seems to think of concepts as defined by giving a simple list of characteristics in no special order; but of all ways of forming concepts, that is one of the least fruitful. If we look through the definitions given in the course of this book, we shall scarcely find one that is of this description. The same is true of the really fruitful definitions in mathematics, such as that of the continuity of a function. What we find in these is not a simple list of characteristics; every element in the definition is intimately, I might almost say organically, connected with the others. A geometrical illustration will make the distinction clear to intuition. If we represent the concepts (or their extensions) by figures or areas in a plane, then the concept defined by a simple list of characteristics corresponds to the area common to all the areas representing the defining characteristics; it is enclosed by segments of their boundary lines. With a definition like this, therefore, what we do—in terms of our illustration—is to use the lines already given in a new way for the purpose of demarcating an area.[2] Nothing essentially new, however, emerges in the process. But the more fruitful type of definition is a matter of drawing boundary lines that were not previously given at all. What we shall be able to infer from it, cannot be inspected in advance; here,

---

[1] On p. 43 [B14] he says that a synthetic proposition can only be seen to be true by the law of contradiction, if another synthetic proposition is pre-supposed.

[2] Similarly, if the characteristics are joined by "or".

nicht einfach aus dem Kasten wieder heraus, was man hinein-
gelegt hatte. Diese Folgerungen erweitern unsere Kenntnisse, und man sollte sie daher Kant zufolge für synthetisch
halten; dennoch können sie rein logisch bewiesen werden
und sind also analytisch. Sie sind in der That in den Definitionen enthalten, aber wie die Pflanze im Samen, nicht
wie der Balken im Hause. Oft braucht man mehre Definitionen zum Beweise eines Satzes, der folglich in keiner
einzelnen enthalten ist und doch aus allen zusammen rein
logisch folgt.

§ 89. Ich muss auch der Allgemeinheit der Behauptung
Kants*) widersprechen: ohne Sinnlichkeit würde uns kein
Gegenstand gegeben werden. Die Null, die Eins sind Gegenstände, die uns nicht sinnlich gegeben werden können. Auch
Diejenigen, welche die kleineren Zahlen für anschaulich halten,
werden doch einräumen müssen, dass ihnen keine der Zahlen,
die grösser als 1000 (1000^{1000}) sind, anschaulich gegeben
werden können, und dass wir dennoch Mancherlei von ihnen
wissen. Vielleicht hat Kant das Wort „Gegenstand" in
etwas anderm Sinne gebraucht; aber dann fallen die Null,
die Eins, unser ∞_1 ganz aus seiner Betrachtung heraus; denn
Begriffe sind sie auch nicht, und auch von Begriffen verlangt
Kant*), dass man ihnen den Gegenstand in der Anschauung
beifüge.

Um nicht den Vorwurf einer kleinlichen Tadelsucht
gegenüber einem Geiste auf mich zu laden, zu dem wir nur
mit dankbarer Bewunderung aufblicken können, glaube ich
auch die Uebereinstimmung hervorheben zu müssen, welche
weit überwiegt. Um nur das hier zunächst Liegende zu
berühren, sehe ich ein grosses Verdienst Kants darin, dass
er die Unterscheidung von synthetischen und analytischen
Urtheilen gemacht hat. Indem er die geometrischen Wahrheiten synthetisch und a priori nannte, hat er ihr wahres

*) A. a. O. III, S. 82.

we are not simply taking out of the box again what we have just put into it. The conclusions we draw from it extend our knowledge, and ought therefore, on KANT's view, to be regarded as synthetic; and yet they can be proved by purely logical means, and are thus analytic. The truth is that they are contained in the definitions, but as plants are contained in their seeds, not as beams are contained in a house. Often we need several definitions for the proof of some proposition, which consequently is not contained in any one of them alone, yet does follow purely logically from all of them together.

§ 89. I must also protest against the generality of KANT's[1] dictum: without sensibility no object would be given to us. Nought and one are objects which cannot be given to us in sensation. And even those who hold that the smaller numbers are intuitable, must at least concede that they cannot be given in intuition any of the numbers greater than 1000 ^{1000} ^{1000}, about which nevertheless we have plenty of information. Perhaps KANT used the word "object" in a rather different sense; but in that case he omits altogether to allow for nought or one, or for our ∞_1,—for these are not concepts either, and even of a concept KANT requires that we should attach its object to it in intuition.

I have no wish to incur the reproach of picking petty quarrels with a genius to whom we must all look up with grateful awe; I feel bound, therefore, to call attention also to the extent of my agreement with him, which far exceeds any disagreement. To touch only upon what is immediately relevant, I consider KANT did great service in drawing the distinction between synthetic and analytic judgements. In calling the truths of geometry synthetic and a priori, he

---

[1] Op. cit., Vol. III, p. 82 [Original edns., A51/B75.]

Wesen enthüllt. Und dies ist noch jetzt werth wiederholt zu werden, weil es noch oft verkannt wird. Wenn Kant sich hinsichtlich der Arithmetik geirrt hat, so thut das, glaube ich, seinem Verdienste keinen wesentlichen Eintrag. Ihm kam es darauf an, dass es synthetische Urtheile a priori giebt; ob sie nur in der Geometrie oder auch in der Arithmetik vorkommen, ist von geringerer Bedeutung.

§ 90. Ich erhebe nicht den Anspruch, die analytische Natur der arithmetischen Sätze mehr als wahrscheinlich gemacht zu haben, weil man immer noch zweifeln kann, ob ihr Beweis ganz aus rein logischen Gesetzen geführt werden könne, ob sich nicht irgendwo ein Beweisgrund andrer Art unvermerkt einmische. Dies Bedenken wird auch durch die Andeutungen nicht vollständig entkräftet, die ich für den Beweis einiger Sätze gegeben habe; es kann nur durch eine lückenlose Schlusskette gehoben werden, sodass kein Schritt geschieht, der nicht einer von wenigen als rein logisch anerkannten Schlussweisen gemäss ist. So ist bis jetzt kaum ein Beweis geführt worden, weil der Mathematiker zufrieden ist, wenn jeder Uebergang zu einem neuen Urtheile als richtig einleuchtet, ohne nach der Natur dieses Einleuchtens zu fragen, ob es logisch oder anschaulich sei. Ein solcher Fortschritt ist oft sehr zusammengesetzt und mehren einfachen Schlüssen gleichwerthig, neben welchen noch aus der Anschauung etwas einfliessen kann. Man geht sprungweise vor, und daraus entsteht die scheinbar überreiche Mannichfaltigkeit der Schlussweisen in der Mathematik; denn je grösser die Sprünge sind, desto vielfachere Combinationen aus einfachen Schlüssen und Anschauungsaxiomen können sie vertreten. Dennoch leuchtet uns ein solcher Uebergang oft umittelbar ein, ohne dass uns die Zwischenstufen zum Bewusstsein kommen, und da er sich nicht als eine der anerkannten logischen Schlussweisen darstellt, sind wir sogleich bereit, dies Einleuchten für ein anschauliches und die erschlossene Wahrheit für eine synthetische zu halten, auch

revealed their true nature. And this is still worth repeating, since even to-day it is often not recognized. If KANT was wrong about arithmetic, that does not seriously detract, in my opinion, from the value of his work. His point was, that there are such things as synthetic judgements a priori; whether they are to be found in geometry only, or in arithmetic as well, is of less importance.

§ 90. I do not claim to have made the analytic character of arithmetical propositions more than probable, because it can still always be doubted whether they are deducible solely from purely logical laws, or whether some other type of premiss is not involved at some point in their proof without our noticing it. This misgiving will not be completely allayed even by the indications I have given of the proof of some of the propositions; it can only be removed by producing a chain of deductions with no link missing, such that no step in it is taken which does not conform to some one of a small number of principles of inference recognized as purely logical. To this day, scarcely one single proof has ever been conducted on these lines; the mathematician rests content if every transition to a fresh judgement is self-evidently correct, without enquiring into the nature of this self-evidence, whether it is logical or intuitive. A single such step is often really a whole compendium, equivalent to several simple inferences, and into it there can still creep along with these some element from intuition. In proofs as we know them, progress is by jumps, which is why the variety of types of inference in mathematics appears to be so excessively rich; for the bigger the jump, the more diverse are the combinations it can represent of simple inferences with axioms derived from intuition. Often, nevertheless, the correctness of such a transition is immediately self-evident to us, without our ever becoming conscious of the subordinate steps condensed within it; whereupon, since it does not obviously conform to any of the recognized types of logical inference, we are prepared to accept its self-evidence forthwith as intuitive, and the conclusion itself as a synthetic

dann, wenn der Geltungsbereich offenbar über das Anschauliche hinausreicht.

Auf diesem Wege ist es nicht möglich, das auf Anschauung beruhende Synthetische von dem Analytischen rein
zu scheiden. Es gelingt so auch nicht, die Axiome der
Anschauung mit Sicherheit vollständig zusammenzustellen,
sodass jeder mathematische Beweis allein aus diesen Axiomen
nach den logischen Gesetzen geführt werden kann.

§ 91. Die Forderung ist also unabweisbar, alle Sprünge
in der Schlussfolgerung zu vermeiden. Dass ihr so schwer
zu genügen ist, liegt an der Langwierigkeit eines schritt-
weisen Vorgehens. Jeder nur etwas verwickeltere Beweis
droht eine ungeheuerliche Länge anzunehmen. Dazu kommt,
dass die übergrosse Mannichfaltigkeit der in der Sprache
ausgeprägten logischen Formen es erschwert, einen Kreis
von Schlussweisen abzugrenzen, der für alle Fälle genügt
und leicht zu übersehen ist.

Um diese Uebelstände zu vermindern, habe ich meine
Begriffsschrift erdacht. Sie soll grössere Kürze und Uebersichtlichkeit des Ausdrucks erzielen und sich in wenigen
festen Formen nach Art einer Rechnung bewegen, sodass
kein Uebergang gestattet wird, der nicht den ein für alle
Mal aufgestellten Regeln gemäss ist*). Es kann sich dann
kein Beweisgrund unbemerkt einschleichen. Ich habe so,
ohne der Anschauung ein Axiom zu entlehnen, einen Satz
bewiesen**), den man beim ersten Blick für einen synthetischen
halten möchte, welchen ich hier so aussprechen will:

Wenn die Beziehung jedes Gliedes einer Reihe zum
nächstfolgenden eindeutig ist, und wenn m und y in dieser
Reihe auf x folgen, so geht y dem m in dieser Reihe vorher
oder fällt mit ihm zusammen oder folgt auf m.

---

*) Sie soll jedoch nicht nur die logische Form wie die boolesche
Bezeichnungsweise, sondern auch einen Inhalt auszudrücken im Stande sein.
**) Begriffsschrift, Halle a/S. 1879, S. 86, Formel 133.

truth—and this even when obviously it holds good of much more than merely what can be intuited.

On these lines what is synthetic and based on intuition cannot be sharply separated from what is analytic. Nor shall we succeed in compiling with certainty a complete set of axioms of intuition, such that from them alone we can derive, by means of the laws of logic, every proof in mathematics.

§ 91. The demand is not to be denied: every jump must be barred from our deductions. That it is so hard to satisfy must be set down to the tediousness of proceeding step by step. Every proof which is even a little complicated threatens to become inordinately long. And moreover, the excessive variety of logical forms that have been developed in our language makes it difficult to isolate a set of modes of inference which is both sufficient to cope with all cases and easy to take in at a glance.

To minimize these drawbacks, I invented my concept writing. It is designed to produce expressions which are shorter and easier to take in, and to be operated like a calculus by means of a small number of standard moves, so that no step is permitted which does not conform to the rules which are laid down once and for all.[1] It is impossible, therefore, for any premiss to creep into a proof without being noticed. In this way I have, without borrowing any axiom from intuition, given a proof of a proposition[2] which might at first sight be taken for synthetic, which I shall here formulate as follows:

If the relation of every member of a series to its successor is (one- or) many-one, and if m and y follow in that series after x, then either y comes in that series before m, or it coincides with m, or it follows after m.

[1] It is designed, however, to be capable of expressing not only the logical form, like Boole's notation, but also the content of a proposition.

[2] *Begriffsschrift*, Halle a/S. 1879, p. 86, Formula 133.

104

Aus diesem Beweise kann man ersehen, dass Sätze, welche unsere Kenntnisse erweitern, analytische Urtheile enthalten können*).

Andere Zahlen.

§ 92. Wir haben unsere Betrachtung bisher auf die Anzahlen beschränkt. Werfen wir nun noch einen Blick auf die andern Zahlengattungen und versuchen wir für dies weitere Feld nutzbar zu machen, was wir auf dem engern erkannt haben!

Um den Sinn der Frage nach der Möglichkeit einer gewissen Zahl klar zu machen, sagt Hankel**):

„Ein Ding, eine Substanz, die selbständig ausserhalb des denkenden Subjects und der sie veranlassenden Objecte existirte, ein selbständiges Princip, wie etwa bei den Pythagoräern, ist die Zahl heute nicht mehr. Die Frage von der Existenz kann daher nur auf das denkende Subject oder die gedachten Objecte, deren Beziehungen die Zahlen darstellen, bezogen werden. Als unmöglich gilt dem Mathematiker streng genommen nur das, was logisch unmöglich ist, d. h. sich selbst widerspricht. Dass in diesem Sinne unmögliche Zahlen nicht zugelassen werden können, bedarf keines Beweises. Sind aber die betreffenden Zahlen logisch möglich, ihr Begriff klar und bestimmt definirt und also

---

*) Diesen Beweis wird man immer noch viel zu weitläufig finden, ein Nachtheil, der vielleicht die fast unbedingte Sicherheit vor einem Fehler oder einer Lücke mehr als aufzuwiegen scheint. Mein Zweck war damals Alles auf die möglichst geringe Zahl von möglichst einfachen logischen Gesetzen zurückzuführen. Infolge dessen wendete ich nur eine einzige Schlussweise an. Ich wies aber schon damals im Vorworte S. VII darauf hin, dass für die weitere Anwendung es sich empfehlen würde, mehr Schlussweisen zuzulassen. Dies kann geschehen ohne der Bündigkeit der Schlusskette zu schaden, und so lässt sich eine bedeutende Abkürzung erreichen.

**) A. a. O. S. 6 u. 7.

From this proof it can be seen that propositions which extend our knowledge can have analytic judgements for their content.[1]

Other numbers.

§ 92. Up to now we have restricted our treatment to the [natural] Numbers. Let us now take a look at the other kinds of numbers, and try to make some use in this wider field of what we have learned in the narrower.

HANKEL,[2] in an attempt to make clear the sense of asking whether some particular type of number is possible, writes as follows:

"Number to-day is no longer a thing, a substance, existing in its own right apart from the thinking subject and the objects which give rise to it, a self-subsistent element in the sort of way it was for the Pythagoreans. The question whether some number exists can therefore only be understood as referring to the thinking subject or to the objects thought about, relations between which the numbers represent. As impossible in the strict sense the mathematician counts only what is logically impossible, that is, self-contradictory. That numbers which are impossible in this sense cannot be admitted, needs no proof. But if the numbers concerned are logically possible, if their concept is clearly and fully defined and there-

---

[1] This proof will certainly still be found far too lengthy, a disadvantage which may, perhaps, be thought to be more than outweighed by the practically absolute certainty that it contains no mistake and no gap. My aim at that time was to reduce everything to the smallest possible number of the simplest possible logical laws. Consequently, I made use of only one principle of deduction. However, even at that time I noted in my Preface, p. vii, that for the further application of my writing it would be imperative to admit more such principles. This can be done without loosening any link in the chain of deduction, and it is possible to achieve in this way a remarkable degree of compression.

[2] Op. cit., pp. 6–7.

ohne Widerspruch, so kann jene Frage nur darauf hinaus-
kommen, ob es im Gebiete des Realen oder des in der Anschauung Wirklichen, des Actuellen ein Substrat derselben,
ob es Objecte gebe, an welchen die Zahlen, also die intellec-
tuellen Beziehungen der bestimmten Art zur Erscheinung
kommen".

§ 93. Bei dem ersten Satze kann man zweifeln, ob
nach Hankel die Zahlen in dem denkenden Subjecte oder
in den sie veranlassenden Objecten oder in beiden existiren.
Im räumlichen Sinne sind sie jedenfalls weder innerhalb noch
ausserhalb weder des Subjects noch eines Objects. Wohl
aber sind sie in dem Sinne ausserhalb des Subjects, dass sie
nicht subjectiv sind. Während jeder nur seinen Schmerz,
seine Lust, seinen Hunger fühlen, seine Ton- und Farben-
empfindungen haben kann, können die Zahlen gemeinsame
Gegenstände für Viele sein, und zwar sind sie für Alle
genau dieselben, nicht nur mehr oder minder ähnliche innere
Zustände von Verschiedenen. Wenn Hankel die Frage von
der Existenz auf das denkende Subject beziehen will, so
scheint er sie damit zu einer psychologischen zu machen, was
sie in keiner Weise ist. Die Mathematik beschäftigt sich
nicht mit der Natur unserer Seele, und wie irgendwelche
psychologische Fragen beantwortet werden, muss für sie völlig
gleichgiltig sein.

§ 94. Auch dass dem Mathematiker nur, was sich
selbst widerspricht, als unmöglich gelte, muss beanstandet
werden. Ein Begriff ist zulässig, auch wenn seine Merkmale
einen Widerspruch enthalten; man darf nur nicht voraussetzen, dass etwas unter ihn falle. Aber daraus, dass der
Begriff keinen Widerspruch enthält, kann noch nicht geschlossen werden, dass etwas unter ihn falle. Wie soll man
übrigens beweisen, dass ein Begriff keinen Widerspruch enthalte? Auf der Hand liegt das keineswegs immer; daraus,
dass man keinen Widerspruch sieht, folgt nicht, dass keiner
da ist, und die Bestimmtheit der Definition leistet keine

fore free from contradiction, then the question whether they exist can amount only to this: Does there exist in reality or in the actual world given to us in intuition a substratum for these numbers, do there exist objects in which they—relations, that is, for the mind, of the type defined—can become phenomenal?"

§ 93. HANKEL's first sentence leaves it doubtful whether he holds that numbers exist in the thinking subject, or in the objects which give rise to them, or in both. In the spatial sense they are, in any case, neither inside nor outside either the subject or any object. But, of course, they are outside the subject in the sense that they are not subjective. Whereas each individual can feel only his own pain or desire or hunger, and can experience only his own sensations of sound and colour, numbers can be objects in common to many individuals, and they are in fact precisely the same for all, not merely more or less similar mental states in different minds. In making the question of the existence of numbers refer to the thinking subject, HANKEL seems to make it a psychological question, which it is not in any way. Mathematics is not concerned with the nature of our mind, and the answer to any question whatsoever in psychology must be for mathematics a matter of complete indifference.

§ 94. Further, exception must be taken to the statement that the mathematician counts as impossible only what is self-contradictory. A concept is still admissible even though its defining characteristics do contain a contradiction: all that we are forbidden to do, is to presuppose that something falls under it. But even if a concept contains no contradiction, we still cannot infer that for that reason something falls under it. If such concepts were not admissible, how could we ever prove that a concept does not contain any contradiction? It is by no means always obvious; it does not follow that because we see no contradiction there is none there, nor does a clear and full definition afford any guarantee against

106

Gewähr dafür. Hankel beweist*), dass ein höheres be-
grenztes complexes Zahlensystem als das gemeine, das allen
Gesetzen der Addition und Multiplication unterworfen wäre,
einen Widerspruch enthält. Das muss eben bewiesen werden;
man sieht es nicht sogleich. Bevor dies geschehen, könnte
immerhin jemand unter Benutzung eines solchen Zahlen-
systems zu wunderbaren Ergebnissen gelangen, deren Begründung nicht schlechter wäre, als die, welche Hankel**)
von den Determinantensätzen mittels der alternirenden Zahlen
giebt; denn wer bürgt dafür, dass nicht auch in deren Begriffe ein versteckter Widerspruch enthalten ist? Und selbst,
wenn man einen solchen allgemein für beliebig viele alter-
nirende Einheiten ausschliessen könnte, würde immer noch
nicht folgen, dass es solche Einheiten gebe. Und grade dies
brauchen wir. Nehmen wir als Beispiel den 18. Satz des 1.
Buches von Euklids Elementen:

In jedem Dreiecke liegt der grössern Seite der grössere
Winkel gegenüber

Um das zu beweisen, trägt Euklid auf der grössern
Seite AC ein Stück AD gleich der kleinern Seite AB ab
und beruft sich dabei auf eine frühere Construction. Der
Beweis würde in sich zusammenfallen, wenn es einen solchen
Punkt nicht gäbe, und es genügt nicht, dass man in dem
Begriffe „Punkt auf AC, dessen Entfernung von A gleich
B ist" keinen Widerspruch entdeckt. Es wird nun B mit
D verbunden. Auch dass es eine solche Gerade giebt, ist
ein Satz, auf den sich der Beweis stüzt.

§ 95. Streng kann die Widerspruchslosigkeit eines
Begriffes wohl nur durch den Nachweis dargelegt werden,
dass etwas unter ihn falle. Das Umgekehrte würde ein
Fehler sein. In diesen verfällt Hankel, wenn er in Bezug
auf die Gleichung x + b = c sagt***):

*) A. a. O. S. 106 u. 107.
**) A. a. O. § 35.
***) A. a. O. S. 5. Aehnlich E. Kossak, a. a. O. S. 17 unten.

it. HANKEL[1] proves that any closed field of complex numbers of higher order than the ordinary, if made subject to all the laws of addition and multiplication, contains a contradiction. Now that is something that needs to be proved; it is not seen immediately. Before his proof was given, anyone could always, by using a number system of that type, have arrived at remarkable results, nor would they have been any worse founded than the theory of determinants, if, with HANKEL,[2] we base that on alternate numbers; for who can assure us that there is not some hidden contradiction in the concept of these numbers also? And moreover, even if we could exclude this possibility generally for as many alternate units as we please, it would still not follow that such units exist. Yet that is precisely what we need. We will take an example from EUCLID's *Elements*, Book I, Theorem 18:

> In any triangle the greater side subtends the greater angle.

To prove this, EUCLID cuts off from the greater side AC a segment AD equal to the lesser side AB, making use for this purpose of a previously given construction. The proof would collapse, if there were no such point as D, and it is not enough that we discover no contradiction in the concept "point on AC whose distance from A is equal to B's". EUCLID proceeds to join BD. That there exists such a line is still another proposition on which the proof depends.

§ 95. Strictly, of course, we can only establish that a concept is free from contradiction by first producing something that falls under it. The converse inference is a fallacy, and one into which HANKEL falls. Referring to [the operation of finding x from] the equation $x + b = c$ [(subtraction)] he says:[3]

---

[1] Op. cit., pp. 106–7.
[2] Op. cit., § 35, pp. 121–4.
[3] Op. cit., p. 5. Similarly E. Kossak, op. cit., p. 17 *ad fin.*

107

„Es liegt auf der Hand, dass es, wenn b $>$ c ist, keine Zahl x in der Reihe 1, 2, 3, . . . giebt, welche die betreffende Aufgabe löst: die Subtraction ist dann unmöglich. Nichts hindert uns jedoch, dass wir in diesem Falle die Differenz (c — b) als ein Zeichen ansehen, welches die Aufgabe löst, und mit welchem genau so zu operiren ist, als wenn es eine numerische Zahl aus der Reihe 1, 2, 3, . . . wäre."

Uns hindert allerdings etwas (2 — 3), ohne Weiteres als Zeichen anzusehen, welches die Aufgabe löst; denn ein leeres Zeichen löst eben die Aufgabe nicht; ohne einen Inhalt ist es nur Tinte oder Druckerschwärze auf Papier, hat als solche physikalische Eigenschaften, aber nicht die, um 3 vermehrt 2 zu geben. Es wäre eigentlich gar kein Zeichen, und sein Gebrauch als solches wäre ein logischer Fehler. Auch in dem Falle, wo c $>$ b, ist nicht das Zeichen („c — b") die Lösung der Aufgabe, sondern dessen Inhalt.

§ 96. Ebensogut könnte man sagen: unter den bisher bekannten Zahlen giebt es keine, welche die beiden Gleichungen

$$x + 1 = 2 \text{ und } x + 2 = 1$$

zugleich befriedigt; aber nichts hindert uns ein Zeichen einzuführen, das die Aufgabe löst. Man wird sagen: die Aufgabe enthält ja einen Widerspruch. Freilich, wenn man als Lösung eine reelle oder gemeine complexe Zahl verlangt; aber erweitern wir doch unser Zahlsystem, schaffen wir doch Zahlen, die den Anforderungen genügen! Warten wir ab, ob uns jemand einen Widerspruch nachweist! Wer kann wissen, was bei diesen neuen Zahlen möglich ist? Die Eindeutigkeit der Subtraction werden wir dann freilich nicht aufrecht erhalten können; aber wir müssen ja auch die Eindeutigkeit des Wurzelziehens aufgeben, wenn wir die negativen Zahlen einführen wollen; durch die complexen Zahlen wird das Logarithmiren vieldeutig.

Schaffen wir auch Zahlen, welche divergirende Reihen zu summiren gestatten! Nein! auch der Mathematiker kann

"It is obvious that, for $b > c$, there is no number x in the series 1, 2, 3, . . . which solves our problem; the subtraction is then *impossible*. There is nothing, however, to prevent us from regarding the difference $(c - b)$ in this case as a *symbol* which solves the problem and which is to be operated with exactly as if it were a figure number in the series 1, 2, 3 . . . ."

Nevertheless, there is something to prevent us from regarding $(2 - 3)$ without more ado as a symbol which solves the problem; for an empty symbol is precisely no solution; without some content it is merely ink or print on paper, as which it possesses physical properties but not that of making 2 when increased by 3. Really, it would not be a symbol at all, and to use it as one would be a mistake in logic. Even for $c > b$, it is not the symbol ("$c - b$") that solves the problem, but its content.

§ 96. We might just as well say this: among numbers hitherto known there is none which satisfies the simultaneous equations

$$x + 1 = 2$$
$$x + 2 = 1,$$

but there is nothing to prevent us from introducing a symbol which solves the problem. Ah, but there is, it will be replied: to satisfy both the equations simultaneously involves a contradiction. Certainly, if we are requiring a real number or an ordinary complex number to satisfy them; but then all we have to do is to widen our number system, to create numbers which do meet these new requirements. Then we can wait and see whether anyone succeeds in producing a contradiction in them. Who can tell what may not be possible with our new numbers? Naturally $(c - b)$ cannot then remain one-valued; but then $\sqrt{a}$ likewise, if we wish to introduce negative numbers, has to cease to be one-valued; and with complex numbers $\log a$ too becomes many-valued.

And why not create still further numbers which permit the summation of diverging series? But that will do,—even

nicht beliebig etwas schaffen, so wenig wie der Geograph; auch er kann nur entdecken, was da ist, und es benennen.

An diesem Irrthum krankt die formale Theorie der Brüche, der negativen, der complexen Zahlen*). Man stellt die Forderung, dass die bekannten Rechnungsregeln für die neu einzuführenden Zahlen möglichst erhalten bleiben, und leitet daraus allgemeine Eigenschaften und Beziehungen ab. Stösst man nirgends auf einen Widerspruch, so hält man die Einführung der neuen Zahlen für gerechtfertigt, als ob ein Widerspruch nicht dennoch irgendwo versteckt sein könnte, und als ob Widerspruchslosigkeit schon Existenz wäre.

§ 97. Dass dieser Fehler so leicht begangen wird, liegt wohl an einer mangelhaften Unterscheidung der Begriffe von den Gegenständen. Nichts hindert uns, den Begriff „Quadratwurzel aus — 1" zu gebrauchen; aber wir sind nicht ohne Weiteres berechtigt, den bestimmten Artikel davor zu setzen und den Ausdruck „die Quadratwurzel aus — 1" als einen sinnvollen anzusehen. Wir können unter der Voraussetzung, dass $i^2 = - 1$ sei, die Formel beweisen, durch welche der Sinus eines Vielfachen des Winkels a durch Sinus und Cosinus von a selbst ausgedrückt wird; aber wir dürfen nicht vergessen, dass der Satz dann die Bedingung $i^2 = - 1$ mit sich führt, welche wir nicht ohne Weiteres weglassen dürfen. Gäbe es gar nichts, dessen Quadrat — 1 wäre, so brauchte die Gleichung kraft unseres Beweises nicht richtig zu sein**), weil die Bedingung $i^2 = - 1$ niemals erfüllt wäre, von der ihre Geltung abhängig erscheint. Es wäre so, als ob wir in einem geometrischen Beweise eine Hilfslinie benutzt hätten, die gar nicht gezogen werden kann.

§ 98. Hankel***) führt zwei Arten von Operationen ein, die er lytische und thetische nennt, und die er durch

---

*) Aehnlich steht es bei Cantors unendlichen Anzahlen.

**) Auf einem andern Wege möchte sie immerhin streng bewiesen werden können.

***) A. a. O. S. 18.

the mathematician cannot create things at will, any more than the geographer can; he too can only discover what is there and give it a name.

This is the error that infects the formalist theory of fractions and of negative and complex numbers.[1] It is made a postulate that the familiar rules of calculation shall still hold, where possible, for the newly-introduced numbers, and from this their general properties and relations are deduced. If no contradiction is anywhere encountered, the introduction of the new numbers is held to be justified, as though it were impossible for a contradiction still to be lurking somewhere nevertheless, and as though freedom from contradiction amounted straight away to existence.

§ 97. That this mistake is so easily made is due, of course, to the failure to distinguish clearly between concepts and objects. Nothing prevents us from using the concept "square root of -1"; but we are not entitled to put the definite article in front of it without more ado and take the expression "the square root of -1" as having a sense. Given that $i^2 = -1$, we can give a proof of the formula expressing the sine of any multiple of the angle a in terms of sin a and cos a; but we ought not to forget that this proposition continues to imply the condition that $i^2 = -1$, which we are not entitled to drop without remark. If there existed nothing at all of which the square was -1, then for all our proof was worth the formula might not be correct,[2] since the condition $i^2 = -1$, on which its validity patently depends, would never be fulfilled. It would be as though in a geometrical proof we had made use of an auxiliary line which is quite impossible to construct.

§ 98. HANKEL[3] introduces two sorts of operation, which he calls lytic and thetic, and which he defines by means of

---

[1] CANTOR's infinite Numbers are in like case.
[2] It might always be possible to prove it strictly in some other way.
[3] Op. cit., p. 18.

gewisse Eigenschaften bestimmt, welche diese Operationen haben sollen. Dagegen ist nichts zu sagen, so lange man nur nicht voraussetzt, dass es solche Operationen und Gegenstände giebt, welche deren Ergebnisse sein können*). Später**) bezeichnet er eine thetische, vollkommen eindeutige, associative Operation durch (a + b) und die entsprechende ebenfalls vollkommen eindeutige lytische durch (a — b). Eine solch-Operation? welche? eine beliebige? dann ist dies keine Definition von (a + b); und wenn es nun keine giebt? Wenn das Wort „Addition" noch keine Bedeutung hätte, wäre es logisch zulässig zu sagen: eine solche Operation wollen wir eine Addition nennen; aber man darf nicht sagen: eine solche Operation soll die Addition heissen und durch (a + b) bezeichnet werden, bevor es feststeht, dass es eine und nur eine einzige giebt. Man darf nicht auf der einen Seite einer Definitionsgleichung den unbestimmten und auf der andern den bestimmten Artikel gebrauchen. Dann sagt Hankel ohne Weiteres: „der Modul der Operation", ohne bewiesen zu haben, dass es einen und nur einen giebt.

§ 99. Kurz diese rein formale Theorie ist unzureichend. Das Werthvolle an ihr ist nur dies. Man beweist, dass wenn Operationen gewisse Eigenschaften wie die Associativität und die Commutativität haben, gewisse Sätze von ihnen gelten Man zeigt nun, dass die Addition und Multiplication, welche man schon kennt, diese Eigenschaften haben, und kann nun sofort jene Sätze von ihnen aussprechen, ohne den Beweis in jedem einzelnen Falle weitläufig zu wiederholen. Erst durch diese Anwendung auf anderweitig gegebene Operationen, gelangt man zu den bekannten Sätzen der Arithmetik. Keineswegs darf man aber glauben die Addition und die Multiplication auf diesem Wege einführen zu können. Man

---

*) Das thut Hankel eigentlich schon durch den Gebrauch der Gleichung Θ (c, b) = a.

**) A. a. O. S. 29.

certain properties that they are to possess. There is nothing against this, so long as it is only not presupposed that operations of these sorts and objects such as their results would be exist.[1] Later[2] he symbolizes an operation which is thetic, one-valued* and associative by $(a + b)$, and the corresponding lytic, and likewise one-valued*, operation by $(a - b)$. *An operation which etc.?* But which one? Any we care to choose? Then that is not a definition of $(a + b)$; and besides, what if none such exists? If the word "addition" had as yet no meaning, it would be quite in order logically to say: we propose to call an operation of this sort an addition; but what we cannot say is: we propose to call an operation of this sort *the* operation of addition, and to symbolize it by $(a + b)$. For it has not yet been established that there is one and only one such operation. We cannot define by putting on one side of our identity the indefinite article and on the other the definite. HANKEL, however, goes on to speak next without more ado of "the modulus of the operation", without having proved that there is one and only one modulus.

§ 99. In a word, this purely formalist theory is not sufficient. What is valuable in it is simply this. We can prove that if any operation possesses certain properties, such as that of being associative or commutative, then certain propositions hold good of it. So that if we go on to show that addition and multiplication, which are already known to us, possess these properties, we can then proceed immediately to assert our propositions of addition and multiplication, without repeating the proof at length for each case individually. Thus it is only after we have applied our formal theory to operations given from elsewhere, that we arrive at the familiar propositions of arithmetic. But we have not the slightest right to suppose that we can use it as a method for introducing addition and multiplication. It does not give

[1] This Hankel really does already by using the identity $\Theta(c, b) = a$.

[2] Op. cit., p. 29.

* [Literally 'perfectly one-valued', a term defined by Hankel.]

giebt nur eine Anleitung für die Definitionen, nicht diese selbst. Man sagt: der Name „Addition" soll nur einer thetischen, vollkommen eindeutigen, associativen Operation gegeben werden, womit diejenige, welche nun so heissen soll, noch gar nicht angegeben ist. Danach stände nichts im Wege, die Multiplication Addition zu nennen und durch (a + b) zu bezeichnen, und niemand könnte mit Bestimmtheit sagen, ob 2 + 3 5 oder 6 wäre.

§ 100. Wenn wir diese rein formale Betrachtungsweise aufgeben, so kann sich aus dem Umstande, dass gleichzeitig mit der Einführung von neuen Zahlen die Bedeutung der Wörter „Summe" und „Product" erweitert wird, ein Weg darzubeiten scheinen. Man nimmt einen Gegenstand, etwa den Mond, und erklärt: der Mond mit sich selbst multiplicirt sei — 1. Dann haben wir in dem Monde eine Quadratwurzel aus — 1. Diese Erklärung scheint gestattet, weil aus der bisherigen Bedeutung der Multiplication der Sinn eines solchen Products noch gar nicht hervorgeht und also bei der Erweiterung dieser Bedeutung beliebig festgesetzt werden kann. Aber wir brauchen auch die Producte einer reelen Zahl mit der Quadratwurzel aus — 1. Wählen wir deshalb lieber den Zeitraum einer Secunde zu einer Quadratwurzel aus — 1 und bezeichnen ihn durch i! Dann werden wir unter 3i den Zeitraum von 3 Secunden verstehen u. s. w.*) Welchen Gegenstand werden wir dann etwa durch 2 + 3i bezeichnen? Welche Bedeutung würde dem Pluszeichen in diesem Falle zu geben sein? Nun das muss

---

*) Mit demselben Rechte könnten wir auch ein gewisses Electricitätsquantum, einen gewissen Flächeninhalt u. s. w. zu Quadratwurzeln aus — 1 wählen, müssten diese verschiedenen Wurzeln dann auch selbstverständlich verschieden bezeichnen. Dass man so beliebig viele Quadratwurzeln aus — 1 scheinbar schaffen kann, wird weniger verwunderlich, wenn man bedenkt, dass die Bedeutung der Quadratwurzel nicht schon vor diesen Festsetzungen unveränderlich feststand, sondern durch sie erst mitbestimmt wird.

their actual definitions, but only lays down the lines for them. We may say: the name "addition" is to be given only to an operation which is thetic, one-valued and associative, but there is nothing at all in this as yet to say which operation it is that is to be so called. So far as this goes, there is nothing to stop us calling multiplication addition and symbolizing it by $(a + b)$, nor could anyone say definitely whether $2 + 3$ was 5 or 6.

§ 100. If we abandon this purely formal method of treatment, we may fasten instead on the circumstance that, simultaneously with the introduction of new numbers, the meanings of the words "sum" and "product" are extended. We take some object, let us say the Moon, and proceed by definition: Let the Moon multiplied by itself be —1. This gives us a square root of —1 in the shape of the Moon. There seems to be nothing wrong with this definition, since the meaning hitherto assigned to multiplication says nothing as to the sense of a product such as the Moon into the Moon, so that as we now come to extend its meaning we can make it, for the Moon, whatever we choose. But we need also the product of a real number into the square root of —1. So let us choose instead as our square root of —1 the time-interval of one second, and let this be symbolized by i. Thus $3i$ will mean the time-interval of 3 seconds, and so on.[1] What object shall we then symbolize by, say, $2 + 3i$? What meaning should be assigned to the plus symbol in this case? Now this must be

[1] We should be equally entitled to choose as further square roots of —1 a certain quantum of electricity, a certain surface area, and so on; but then we should naturally have to use different symbols to signify these different roots. That we are able, apparently, to create in this way as many square roots of —1 as we please, is not so astonishing when we reflect that the meaning of the square root of —1 is not something which was already unalterably fixed before we made these choices, but is decided for the first time by and along with them.

allgemein festgesetzt werden, was freilich nicht leicht sein wird. Doch nehmen wir einmal an, dass wir allen Zeichen von der Form a + bi einen Sinn gesichert hätten, und zwa einen solchen, dass die bekannten Additionssätze gelten! Dann müssten wir ferner festsetzen, dass allgemein

$$(a + bi)(c + di) = ac - bd + i(ad + bc)$$

sein solle, wodurch wir die Multiplication weiter bestimmen würden.

§ 101. Nun könnten wir die Formel für cos (n a) beweisen, wenn wir wüssten, dass aus der Gleichheit complexer Zahlen die Gleichheit der reellen Theile folgt. Das müsste aus dem Sinne von a + bi hervorgehn, den wir hier als vorhanden angenommen haben. Der Beweis würde nur für den Sinn der complexen Zahlen, ihrer Summen und Producte gelten, den wir festgesetzt haben. Da nun für ganzes reelles n und reelles a i gar nicht mehr in der Gleichung vorkommt, so ist man versucht zu schliessen: also ist es ganz gleichgiltig, ob i eine Secunde, ein Millimeter oder was sonst bedeutet, wenn nur unsere Additions- und Multiplicationssätze gelten; auf die allein kommt es an; um das Uebrige brauchen wir uns nicht zu kümmern. Vielleicht kann man die Bedeutung von a + bi, von Summe und Product in verschiedener Weise so festsetzen, dass jene Sätze bestehen bleiben; aber es ist nicht gleichgiltig, ob man überhaupt einen solchen Sinn für diese Ausdrücke finden kann.

§ 102. Man thut oft so, als ob die blosse Forderung schon ihre Erfüllung wäre. Man fordert, dass die Subtraction*), die Division, die Radicirung immer ausführbar seien, und glaubt damit genug gethan zu haben. Warum fordert man nicht auch, dass durch beliebige drei Punkte eine Gerade gezogen werde? Warum fordert man nicht, dass für ein dreidimensionales complexes Zahlensystem sämmtliche Addi-

---

*) Vergl. Kossak a. a. O. S. 17.

laid down generally for all such cases, which clearly is not going to be easy. However, let us just assume that we have successfully secured a sense for all symbols of the form $a + bi$, and a sense such that the familiar laws of addition hold good of it. What we should then have to do would be to lay it down further that in general

$$(a + bi)(c + di) = ac - bd + i(ad + bc),$$

thus defining the extended meaning of multiplication.

§ 101. We should now be able to prove the formula for cos $(n\,a)$, if we knew that from the identity of complex numbers the identity of their real parts can be inferred. That would have to result from the sense of $a + bi$, which we are here taking to have been made available. Our proof of the formula would thus hold only for complex numbers and their sums and products in the sense fixed by us. Now since for real integral n and real a i disappears completely from the identity in the formula, we are tempted to conclude therefore that it is quite immaterial whether i means a second or a milli-metre or anything else, provided only that our laws of addition and multiplication hold good; everything depends on that, and the rest we need not bother about. Well, perhaps it is indeed possible to assign a whole variety of different meanings to $a + bi$, and to sum and product, all of them such that those laws continue to hold good; but it is not immaterial whether we can or cannot find *some* such a sense for those expressions.

§ 102. It is common to proceed as if a mere postulation were equivalent to its own fulfilment. We postulate that it shall be possible in all cases to carry out the operation of subtraction,[1] or of division, or of root extraction, and suppose that with that we have done enough. But why do we not postulate that through any three points it shall be possible to draw a straight line? Why do we not postulate that all the laws

[1] Cp. Kossak, op. cit., p. 17.

tions- und Multiplicationssätze gelten wie für ein reelles? Weil diese Forderung einen Widerspruch enthält. Ei so beweise man denn erst, dass jene andern Forderungen keinen Widerspruch enthalten! Ehe man das gethan hat, ist alle vielerstrebte Strenge nichts als eitel Schein und Dunst.

In einem geometrischen Lehrsatze kommt die zum Beweise etwa gezogene Hilfslinie nicht vor. Vielleicht sind mehre möglich z. B., wenn man einen Punkt willkührlich wählen kann. Aber wie entbehrlich auch jede einzelne sein mag, so hängt doch die Beweiskraft daran, dass man eine Linie von der verlangten Beschaffenheit ziehen könne. Die blosse Forderung genügt nicht. So ist es auch in unserm Falle für die Beweiskraft nicht gleichgiltig, ob „a + bi" einen Sinn hat oder blosse Druckerschwärze ist. Es reicht dazu nicht hin, zu verlangen, es solle einen Sinn haben, oder zu sagen, der Sinn sei die Summe von a und bi, wenn man nicht vorher erklärt hat, was „Summe" in diesem Falle bedeutet, und wenn man den Gebrauch des bestimmten Artikels nicht gerechtfertigt hat.

§ 103. Gegen die von uns versuchte Festsetzung des Sinnes von „i" lässt sich freilich Manches einwenden. Wir bringen dadurch etwas ganz Fremdartiges, die Zeit, in die Arithmetik. Die Secunde steht in gar keiner innern Beziehung zu den reellen Zahlen. Die Sätze, welche mittels der complexen Zahlen bewiesen werden, würden Urtheile a posteriori oder doch synthetische sein, wenn es keine andere Art des Beweises gäbe, oder wenn man für i keinen andern Sinn finden könnte. Zunächst muss jedenfalls der Versuch gemacht werden, alle Sätze der Arithmetik als analytische nachzuweisen.

Wenn Kossak*) in Bezug auf die complexe Zahl sagt: „Sie ist die zusammengesetzte Vorstellung von ver-

*) A. a. O. S. 17.

of addition and multiplication shall continue to hold for a three-dimensional complex number system just as they do for real numbers? Because this postulate contains a contradiction. Very well then, what we have to do first is to prove that these other postulates of ours do not contain any contradiction. Until we have done that, all rigour, strive for it as we will, is so much moonshine.

In a geometrical theorem where a constructed line is used for the proof, the auxiliary line does not occur in the theorem. Perhaps more than one such line is possible, as for instance where we can select a point at will. But however much we can dispense with each and any of them individually, still the cogency of our proof depends on its being possible to draw some line of the required character. Merely to postulate it is not enough. So in our case likewise, it is not immaterial to the cogency of our proof whether "$a + bi$" has a sense or is nothing more than printer's ink. It will not get us anywhere simply to require that it have a sense, or to say that it is to have the sense of the sum of a and bi, when we have not previously defined what "sum" means in this case and when we have given no justification for the use of the definite article.

§ 103. Against the particular sense we have proposed to assign to "i" many objections can of course be brought. By it, we are importing into arithmetic something quite foreign to it, namely time. The second stands in absolutely no intrinsic relation to the real numbers. Propositions proved by the aid of complex numbers would become a posteriori judgements, or rather, at any rate, synthetic, unless we could find some other sort of proof for them or some other sense for i. We must at least first make the attempt to show that all propositions of arithmetic are analytic.

KOSSAK's[1] account of complex number—"the composite

---

[1] Op. cit., p. 17.

schiedenartigen Gruppen unter einander gleicher Elemente*)",
so scheint er damit die Einmischung von Fremdartigem vermieden zu haben; aber er scheint es auch nur infolge der
Unbestimmtheit des Ausdrucks. Man erhält gar keine Antwort darauf, was 1 + i eigentlich bedeute: die Vorstellung
eines Apfels und einer Birne oder die von Zahnweh und
Podagra? Beide zugleich kann es doch nicht bedeuten, weil
dann 1 + i nicht immer gleich 1 + i wäre. Man wird sagen:
das kommt auf die besondere Festsetzung an. Nun, dann
haben wir auch in K o s s a k 's Satze noch gar keine Definition
der complexen Zahl, sondern nur eine allgemeine Anleitung
dazu. Wir brauchen aber mehr; wir müssen bestimmt wissen,
was „i" bedeutet, und wenn wir nun jener Anleitung folgend
sagen wollten: die Vorstellung einer Birne, so würden wir
wieder etwas Fremdartiges in die Arithmetik einführen.

Das, was man die geometrische Darstellung complexer
Zahlen zu nennen pflegt, hat wenigstens den Vorzug vor
den bisher betrachteten Versuchen, dass dabei 1 und i nicht
ganz ohne Zusammenhang und ungleichartig erscheinen
sondern dass die Strecke, welche man als Darstellung von i
betrachtet, in einer gesetzmässigen Beziehung zu der Strecke
steht, durch welche 1 dargestellt wird. Uebrigens ist es
genau genommen nicht richtig, dass hierbei 1 eine gewisse
Strecke, i eine zu ihr senkrechte von gleicher Länge bedeute,
sondern „1" bedeutet überall dasselbe. Eine complexe Zahl
giebt hier an, wie die Strecke, welche als ihre Darstellung
gilt, aus einer gegebenen Strecke (Einheitsstrecke) durch
Vervielfältigung, Theilung und Drehung**) hervorgeht. Aber
auch hiernach erscheint jeder Lehrsatz, dessen Beweis sich
auf die Existenz einer complexen Zahl stützen muss, von
der geometrischen Anschauung abhängig und also synthetisch.

---

*) Man vergleiche über den Ausdruck „Vorstellung" § 27, über
„Gruppe" das in Bezug auf „Aggregat" § 23 u. § 25 Gesagte, über die Gleich-
heit der Elemente §§ 34–39.
**) Der Einfachheit wegen sehe ich hier vom Incommensurabeln ab.

idea of heterogeneous groups of identical elements"[1]—appears to avoid importing anything foreign, but this appearance is only due to the vagueness of his terminology. We are given no answer at all to the question, what does $1 + i$ really mean? Is it the idea of an apple and a pear, or the idea of toothache and gout? Not both at once, at any rate, because then $1 + i$ would not be always identical with $1 + i$. The temptation is to say: it depends on the special meaning we assign to it. Very well then, KOSSAK's statement once again does not yet give us any definition at all of complex number, it only lays down the general lines to proceed along. But we need more; we must know definitely what "i" means, and if we do proceed along his lines and try saying it means the idea of a pear, we shall once again be introducing something foreign into arithmetic.

What is commonly called the geometrical representation of complex numbers has at least this advantage over the proposals so far considered, that in it 1 and i do not appear as wholly unconnected and different in kind: the segment taken to represent i stands in a regular relation to the segment which represents 1. Though I may add that, strictly, it is not correct that 1 here means a certain segment and i a segment perpendicular to it of the same length; on the contrary, "1" means in all contexts the same. A complex number, on this interpretation, shows how the segment taken as its representation is reached, starting from a given segment (the unit segment), by means of operations of multiplication, division, and rotation.[2] However, even this account seems to make every theorem whose proof has to be based on the existence of a complex number dependent on geometrical intuition and so synthetic.

[1] Cf. for the term "idea" § 27, for "group" what is said about "agglomeration" in § 23 and § 25, and for the identity of the elements §§ 34–39.
[2] For simplicity I neglect incommensurables here.

114

§ 104. Wodurch sollen uns denn nun die Brüche, die Irrationalzahlen und die complexen Zahlen gegeben werden? Wenn wir die Anschauung zu Hilfe nehmen, so führen wir etwas Fremdartiges in die Arithmetik ein; wenn wir aber nur den Begriff einer solchen Zahl durch Merkmale bestimmen, wenn wir nur verlangen, dass die Zahl gewisse Eigenschaften habe, so bürgt nichts dafür, dass auch etwas unter den Begriff falle und unsern Anforderungen entspreche, und doch müssen sich grade hierauf Beweise stützen.

Nun, wie ist es denn bei der Anzahl? Dürfen wir wirklich von 1000 (1000^{1000}) nicht reden, bevor uns nicht soviele Gegenstände in der Anschauung gegeben sind? Ist es so lange ein leeres Zeichen? Nein! es hat einen ganz bestimmten Sinn, obwohl es psychologisch schon in Anbetracht der Kürze unseres Lebens unmöglich ist, uns soviele Gegenstände vor das Bewusstsein zu führen*); aber trotzdem ist 1000 (1000^{1000}) ein Gegenstand, dessen Eigenschaften wir erkennen können, obgleich er nicht anschaulich ist. Man überzeugt sich davon, indem man bei der Einführung des Zeichens a^n für die Potenz zeigt, dass immer eine und nur eine positive ganze Zahl dadurch ausgedrückt wird, wenn a und n positive ganze Zahlen sind. Wie dies geschehen kann, würde hier zu weit führen, im Einzelnen darzulegen. Die Weise, wie wir im § 74 die Null, in § 77 die Eins, in § 84 die unendliche Anzahl ∞_1 erklärt haben, und die Andeutung des Beweises, dass auf jede endliche Anzahl in der natürlichen Zahlenreihe eine Anzahl unmittelbar folgt (§§ 82 u. 83), werden den Weg im Allgemeinen erkennen lassen.

Es wird zuletzt auch bei der Definition der Brüche, complexen Zahlen u. s. w. Alles darauf ankommen, einen beurtheilbaren Inhalt aufzusuchen, der in eine Gleichung verwandelt werden kann, deren Seiten dann eben die neuen

---

*) Ein leichter Ueberschlag zeigt, dass dazu Millionen Jahre lange nicht hinreichen würden.

§ 104. How are complex numbers to be given to us then, and fractions and irrational numbers? If we turn for assistance to intuition, we import something foreign into arithmetic; but if we only define the concept of such a number by giving its characteristics, if we simply require the number to have certain properties, then there is still no guarantee that anything falls under the concept and answers to our requirements, and yet it is precisely on this that proofs must be based.

Well, how do things stand with the [natural] Numbers? Have we really no right to speak of $1000^{1000^{1000}}$ until such time as that many objects have been given to us in intuition? Is it, till then, an empty symbol? Not at all. It has a perfectly definite sense, even although, psychologically speaking and having regard to the shortness of human life, it is impossible for us ever to become conscious of that many objects;[1] in spite of that, $1000^{1000^{1000}}$ is still an object, whose properties we can come to know, even though it is not intuitable. To convince ourselves of this, we have only to show, introducing the symbol a^n for the nth power of a, that for positive integral a and n this expression always refers to one and only one positive whole number. To give the proof of this in detail would take us too far afield for present purposes. A general idea of the way it goes can be gathered from the method used to define nought in § 74, one in § 77, and the infinite Number ∞_1 in § 84, and from the outline of the proof that after every finite Number in the series of natural numbers a Number directly follows (§§ 82-3).

In the same way with the definitions of fractions, complex numbers and the rest, everything will in the end come down to the search for a judgement-content which can be transformed into an identity whose sides precisely are the new

[1] A simple calculation shows that millions of years would not be time enough for that.

115

Zahlen sind. Mit andern Worten: wir müssen den Sinn eines Wiedererkennungsurtheils für solche Zahlen festsetzen. Dabei sind die Bedenken zu beachten, die wir (§§ 63—68) in Betreff einer solchen Umwandlung erörtert haben. Wenn wir ebenso wie dort verfahren, so werden uns die neuen Zahlen als Umfänge von Begriffen gegeben.

§ 105. Aus dieser Auffassung der Zahlen*) erklärt sich, wie mir scheint, leicht der Reiz, den die Beschäftigung mit der Arithmetik und Analysis ausübt. Man könnte wohl mit Abänderung eines bekannten Satzes sagen: der eigentliche Gegenstand der Vernunft ist die Vernunft. Wir beschäftigen uns in der Arithmetik mit Gegenständen, die uns nicht als etwas Fremdes von aussen durch Vermittelung der Sinne bekannt werden, sondern die unmittelbar der Vernunft gegeben sind, welche sie als ihr Eigenstes völlig durchschauen kann**).

Und doch, oder vielmehr grade daher sind diese Gegenstände nicht subjective Hirngespinnste. Es giebt nichts Objectiveres als die arithmetischen Gesetze.

§ 106. Werfen wir noch einen kurzen Rückblick auf den Gang unserer Untersuchung! Nachdem wir festgestellt hatten, dass die Zahl weder ein Haufe von Dingen noch eine Eigenschaft eines solchen, dass sie aber auch nicht subjectives Erzeugniss seelischer Vorgänge ist; sondern dass die Zahlangabe von einem Begriffe etwas Objectives aussage, versuchten wir zunächst die einzelnen Zahlen o, 1 u. s. w. und das Fortschreiten in der Zahlenreihe zu definiren. Der erste Versuch misslang, weil wir nur jene Aussage von

---

*) Man könnte sie auch formal nennen. Doch ist sie ganz verschieden von der oben unter diesem Namen beurtheilten.

**) Ich will hiermit gar nicht leugnen, dass wir ohne sinnliche Eindrücke dumm wie ein Brett wären und weder von Zahlen noch von sonst etwas wüssten; aber dieser psychologische Satz geht uns hier gar nichts an. Wegen der beständigen Gefahr der Vermischung zweier grundverschiedener Fragen hebe ich dies nochmals hervor.

numbers. In other words, what we must do is fix the sense of a recognition-judgement for the case of these numbers. In doing so, we must not forget the doubts raised by such transformations, which we discussed in §§ 63–68. If we follow the same procedure as we did there, then the new numbers are given to us as extensions of concepts.

§ 105. On this view of numbers[1] the charm of work on arithmetic and analysis is, it seems to me, easily accounted for. We might say, indeed, almost in the well-known words: the reason's proper study is itself. In arithmetic we are not concerned with objects which we come to know as something alien from without through the medium of the senses, but with objects given directly to our reason and, as its nearest kin, utterly transparent to it.[2]

And yet, or rather for that very reason, these objects are not subjective fantasies. There is nothing more objective than the laws of arithmetic.

§ 106. Let us cast a final brief glance back over the course of our enquiry. After establishing that number is neither a collection of things nor a property of such, yet at the same time is not a subjective product of mental processes either, we concluded that a statement of number asserts something objective of a concept. We attempted next to define the individual numbers 0, 1, etc., and the step from one number to the next in the number series. Our first attempt broke down, because we had defined only the predicate which we said was

---

[1] It too might be called formalist. However, it is completely different from the view criticized above under that name.

[2] By this I do not mean in the least to deny that without sense impressions we should be as stupid as stones, and should know nothing either of numbers or of anything else; but this psychological proposition is not of the slightest concern to us here. Because of the ever-present danger of confusing two fundamentally different questions, I make this point once more.

116

Begriffen, nicht aber die o, die 1 abgesondert definirt hatten, welche nur Theile von ihr sind. Dies hatte zur Folge, dass wir die Gleichheit von Zahlen nicht beweisen konnten. Es zeigte sich, dass die Zahl, mit der sich die Arithmetik beschäftigt, nicht als ein unselbständiges Attribut, sondern substantivisch gefasst werden muss*). Die Zahl erschien so als wiedererkennbarer Gegenstand, wenn auch nicht als physikalischer oder auch nur räumlicher noch als einer, von dem wir uns durch die Einbildungskraft ein Bild entwerfen können. Wir stellten nun den Grundsatz auf, dass die Bedeutung eines Wortes nicht vereinzelt, sondern im Zusammenhange eines Satzes zu erklären sei, durch dessen Befolgung allein, wie ich glaube, die physikalische Auffassung der Zahl vermieden werden kann, ohne in die psychologische zu verfallen. Es giebt nun eine Art von Sätzen, die für jeden Gegenstand einen Sinn haben müssen, das sind die Wiedererkennungsätze, bei den Zahlen Gleichungen genannt. Auch die Zahlangabe, sahen wir, ist als eine Gleichung aufzufassen. Es kam also darauf an, den Sinn einer Zahlengleichung festzustellen, ihn auszudrücken, ohne von den Zahlwörtern oder dem Worte „Zahl" Gebrauch zu machen. Die Möglichkeit, die unter einen Begriff F fallenden Gegenstände den unter einen Begriff G fallenden beiderseits eindeutig zuzuordnen, erkannten wir als Inhalt eines Wiedererkennungsurtheils von Zahlen. Unsere Definition musste also jene Möglichkeit als gleichbedeutend mit einer Zahlengleichung hinstellen. Wir erinnerten an ähnliche Fälle: die Definition der Richtung aus dem Parallelismus, der Gestalt aus der Aehnlichkeit u. s. w.

§ 107. Es erhob sich nun die Frage: wann ist man berechtigt, einen Inhalt als den eines Wiedererkennungsurtheils aufzufassen? Es muss dazu die Bedingung erfüllt

---

*) Der Unterschied entspricht dem zwischen „blau" und „die Farbe des Himmels".

asserted of the concept, but had not given separate definitions of o or 1, which are only elements in such predicates. This resulted in our being unable to prove the identity of numbers. It became clear that the number studied by arithmetic must be conceived not as a dependent attribute, but substantivally.[1] Number thus emerged as an object that can be recognized again, although not as a physical or even a merely spatial object, nor yet as one of which we can form a picture by means of our imagination. We next laid down the fundamental principle that we must never try to define the meaning of a word in isolation, but only as it is used in the context of a proposition: only by adhering to this can we, as I believe, avoid a physical view of number without slipping into a psychological view of it. Now for every object there is one type of proposition which must have a sense, namely the recognition-statement, which in the case of numbers is called an identity. Statements of number too are, we saw, to be considered as identities. The problem, therefore, was now this: to fix the sense of a numerical identity, that is, to express that sense without making use of number words or the word "number". The content of a recognition-judgement concerning numbers we found to be this, that it is possible to correlate one to one the objects falling under a concept F with those falling under a concept G. Accordingly, our definition had to lay it down that a statement of this possibility means the same as a numerical identity. We recalled similar cases: the definition of direction derived from parallelism of lines, that of shape derived from similarity of figures, and so on.

§ 107. The question then arose: when are we entitled to regard a content as that of a recognition-judgement? For this a certain condition has to be satisfied, namely that it

---

[1] The distinction corresponds to that between "blue" and "the colour of the sky".

sein, dass in jedem Urtheile unbeschadet seiner Wahrheit die linke Seite der versuchsweise angenommenen Gleichung durch die rechte ersetzt werden könne. Nun ist uns, ohne dass weitere Definitionen hinzukommen, zunächst von der linken oder rechten Seite einer solchen Gleichung keine Aussage weiter bekannt als eben die der Gleichheit. Es brauchte also die Ersetzbarkeit nur in einer Gleichung nachgewiesen zu werden.

Aber es blieb noch ein Bedenken bestehen. Ein Wiedererkennungssatz muss nämlich immer einen Sinn haben. Wenn wir nun die Möglichkeit, die unter den Begriff F fallenden Gegenstände den unter den Begriff G fallenden beiderseits eindeutig zuzuordnen, als eine Gleichung auffassen, indem wir dafür sagen: „die Anzahl, welche dem Begriffe F zukommt, ist gleich der Anzahl, welche dem Begriffe G zukommt," und hiermit den Ausdruck „die Anzahl, welche dem Begriffe F zukommt" einführen, so haben wir für die Gleichung nur dann einen Sinn, wenn beide Seiten die eben genannte Form haben. Wir könnten nach einer solchen Definition nicht beurtheilen, ob˙eine Gleichung wahr oder falsch ist, wenn nur die eine Seite diese Form hat. Das veranlasste uns zu der Definition:

Die Anzahl, welche dem Begriffe F zukommt, ist der Umfang des Begriffes „Begriff gleichzahlig dem Begriffe F", indem wir einen Begriff F gleichzahlig einem Begriffe G nannten, wenn jene Möglichkeit der beiderseits eindeutigen Zuordnung besteht.

Hierbei setzten wir den Sinn des Ausdruckes „Umfang des Begriffes" als˗ bekannt voraus. Diese Weise, die Schwierigkeit zu überwinden, wird wohl nicht überall Beifall finden, und Manche werden vorziehn, jenes Bedenken in andrer Weise zu beseitigen. Ich lege auch auf die Heranziehung des Umfangs eines Begriffes kein entscheidendes Gewicht.

§ 108. Es blieb nun noch übrig die beiderseits eindeutige Zuordnung zu erklären; wir führten sie auf rein

must be possible in every judgement to substitute without loss of truth the right-hand side of our putative identity for its left-hand side. Now at the outset, and until we bring in further definitions, we do not know of any other assertion concerning either side of such an identity except the one, that they are identical. We had only to show, therefore, that the substitution is possible in an identity.

One doubt, however, still remained, which was this. A recognition-statement must always have a sense. But now if we treat the possibility of correlating one to one the objects falling under the concept F with the objects falling under the concept G as an identity, by putting for it: "the Number which belongs to the concept F is identical with the Number which belongs to the concept G", thus introducing the expression "the Number which belongs to the concept F", this gives us a sense for the identity only if both sides of it are of the form just mentioned. A definition like this is not enough to enable us to decide whether an identity is true or false if only one side of it is of this form. We were thus led to give the definition:

The Number which belongs to the concept F is the extension of the concept "concept equal to the concept F", where a concept F is called equal to a concept G if there exists the possibility of one-one correlation referred to above.

In this definition the sense of the expression "extension of a concept" is assumed to be known. This way of getting over the difficulty cannot be expected to meet with universal approval, and many will prefer other methods of removing the doubt in question. I attach no decisive importance even to bringing in the extensions of concepts at all.

§ 108. It now still remained to define one-one correlation; this we reduced to purely logical relationships. Next, we

logische Verhältnisse zurück. Nachdem wir nun den Beweis des Satzes angedeutet hatten: die Zahl, welche dem Begriffe F zukommt, ist gleich der, welche dem Begriffe G zukommt, wenn der Begriff F dem Begriffe G gleichzahlig ist, definirten wir die o, den Ausdruck „n folgt in der natürlichen Zahlenreihe unmittelbar auf m" und die Zahl 1 und zeigten, dass 1 in der natürlichen Zahlenreihe unmittelbar auf o folgt. Wir führten einige Sätze an, die sich an dieser Stelle leicht beweisen lassen, und gingen dann etwas näher auf folgenden ein, der die Unendlichkeit der Zahlenreihe erkennen lässt:

Auf jede Zahl folgt in der natürlichen Zahlenreihe eine Zahl.

Wir wurden hierdurch auf den Begriff „der mit n endenden natürlichen Zahlenreihe angehörend" geführt, von dem wir zeigen wollten, dass die ihm zukommende Anzahl auf n in der natürlichen Zahlenreihe unmittelbar folge. Wir definirten ihn zunächst mittels des Folgens eines Gegenstandes y auf einen Gegenstand x in einer allgemeinen φ - Reihe. Auch der Sinn dieses Ausdruckes wurde auf rein logische Verhältnisse zurückgeführt. Und dadurch gelang es, die Schlussweise von n auf (n + 1), welche gewöhnlich für eine eigenthümlich mathematische gehalten wird, als auf den allgemeinen logischen Schlussweisen beruhend nachzuweisen.

Wir brauchten nun zum Beweise der Unendlichkeit der Zahlenreihe den Satz, dass keine endliche Zahl in der natürlichen Zahlenreihe auf sich selber folgt. Wir kamen so zu den Begriffen der endlichen und der unendlichen Zahl. Wir zeigten, dass der letztere im Grunde nicht weniger logisch gerechtfertigt als der erstere ist. Zum Vergleiche wurden Cantors unendliche Anzahlen und dessen „Folgen in der Succession" herangezogen, wobei auf die Verschiedenheit im Ausdrucke hingewiesen wurde.

§ 109. Aus allem Vorangehenden ergab sich nun mit grosser Wahrscheinlichkeit die analytische und apriorische Natur der arithmetischen Wahrheiten; und wir gelangten

first gave an outline of the proof of the proposition: the number which belongs to the concept F is identical with the number which belongs to the concept G, if the concept F is equal to the concept G; and then gave definitions of nought, of the expression "n follows in the series of natural numbers directly after m", and of the number 1, showing that 1 follows in the series of natural numbers directly after 0. After adducing a number of propositions which can easily be proved at this stage, we proceeded to go rather more closely into the following proposition, from which we learn that the number series is infinite:

After every number there follows in the series of natural numbers a number.

This led us to the concept "member of the series of natural numbers ending with n", with the aim of showing that the Number belonging to this concept follows in the series of natural numbers directly after n. We began by defining this in terms of the following of an object y after an object x in a series in general ϕ. The sense of this expression too was reduced to purely logical relationships. And by this means we succeeded in showing that the inference from n to $(n+1)$, which is ordinarily held to be peculiar to mathematics, is really based on the universal principles of inference in logic.

Now to prove that the number series is infinite, we needed to make use of the proposition that no finite number follows in the series of natural numbers after itself. And so we arrived at the concepts of finite and infinite number. We showed that fundamentally the latter is no less logically justified than the former. For the sake of comparison, CANTOR's infinite Numbers and his "following in the succession" were referred to, and at the same time the divergence in his terminology was pointed out.

§ 109. From all the preceding it thus emerged as a very probable conclusion that the truths of arithmetic are analytic and a priori; and we achieved an improvement on the view of

zu einer Verbesserung der Ansicht Kants. Wir sahen ferner, was noch fehlt, um jene Wahrscheinlichkeit zur Gewissheit zu erheben, und gaben den Weg an, der dahin führen muss.

Endlich benutzten wir unsere Ergebnisse zur Kritik einer formalen Theorie der negativen, gebrochenen, irrationalen und complexen Zahlen, durch welche deren Unzulänglichkeit offenbar wurde. Ihren Fehler erkannten wir darin, dass sie die Widerspruchslosigkeit eines Begriffes als bewiesen annahm, wenn sich kein Widerspruch gezeigt hatte, and dass die Widerspruchslosigkeit eines Begriffes schon als hinreichende Gewähr für seine Erfülltheit galt. Diese Theorie bildet sich ein, sie brauche nur Forderungen zu stellen; deren Erfüllung verstehe sich dann von selbst. Sie gebärdet sich wie ein Gott, der durch sein blosses Wort schaffen kann, wessen er bedarf. Es musste auch gerügt werden, wenn eine Anweisung zur Definition für diese selbst ausgegeben wurde, eine Anweisung, deren Befolgung Fremdartiges in die Arithmetik einführen würde, obwohl sie selbst im Ausdrucke sich davon frei zu halten vermag, aber nur weil sie blosse Anweisung bleibt.

So geräth jene formale Theorie in Gefahr, auf das Aposteriorische oder doch Synthetische zurückzufallen, wie sehr sie sich auch den Anschein giebt, in der Höhe der Abstractionen zu schweben.

Unsere frühere Betrachtung der positiven ganzen Zahlen zeigte uns nun die Möglichkeit, die Einmischung von äussern Dingen und geometrischen Anschauungen zu vermeiden, ohne doch in den Fehler jener formalen Theorie zu verfallen. Es kommt wie dort darauf an, den Inhalt eines Wiedererkennungsurtheils festzusetzen. Denken wir dies überall geschehen, so erscheinen die negativen, gebrochenen, irrationalen und complexen Zahlen nicht geheimnissvoller als die positiven ganzen Zahlen, diese nicht reeller, wirklicher, greifbarer als jene.

KANT. We saw further what is still needed to raise this probability to a certainty, and indicated the path which must lead to that goal.

Finally, we made use of our results in a critique of a formalist theory of negative, fractional, irrational and complex numbers, which made the inadequacy of the theory evident. We came to see that its error lies in taking it as proved that a concept is free from contradiction if no contradiction has revealed itself, and in taking freedom from contradiction in a concept as sufficient guarantee in itself that something falls under it. This theory imagines that all we need do is make postulates; that these are satisfied then goes without saying. It conducts itself like a god, who can create by his mere word whatever he wants. It had also to be censured for passing off as a definition what is only a guide towards a definition, and one which, if we followed it, would lead to the introduction into arithmetic of foreign elements; these do not, it is true, obtrude into the words of the "definition", but only because it remains a mere guide.

The formalists are thus in danger of relapsing into an a posteriori or at any rate a synthetic theory, however high on the summits of abstraction they may seem to themselves to be floating.

Now we, from our previous treatment of the positive whole numbers, have seen that it is possible to avoid all importation of external things and geometrical intuitions into arithmetic, without, for all that, falling into the error of the formalists. Here, just as there, it is a matter of fixing the content of a recognition-judgement. Once suppose this everywhere accomplished, and numbers of every kind, whether negative, fractional, irrational or complex, are revealed as no more mysterious than the positive whole numbers, which in turn are no more real or more actual or more palpable than they.